FLUVIAL GEOMORPHOLOGY OF SAPT KOSI RIVER BASIN

Dr. Rajesh Kumar Mahato
Assistant Teacher (PGT), Geography
N.D. Rashtriya Vidyalaya (H.S.)
Sitarampur, Paschim Burdwan, West Bengal, India

RAJESH PUBLICATIONS
NEW DELHI - 110002

Cover Photo Credit : Author
Front Page Upper Photo : Indrawaty River channel bed near Dolaghat, Nepal
Front Page Lower Photo : Tamba Kosi river channel, Charikot, Nepal
Back Page Photo : Sapt Kosi river enters in Terai region, Chatra, Nepal

ISBN : 978-93-83684-55-7

Published by :
Sanjay Gupta
RAJESH PUBLICATIONS
1, Ansari Road, Darya Ganj
New Delhi - 110002
Ph.: 011-23274550

Printed in India by :
Vijeta Offset Printers
Delhi-110002

First Published in India 2018

Price : Rs. 800.00

Dedicated to

All living organisms, loose their life (almost every year) during peak fluvial monsoon flood over Sapt Kosi Plain of Bihar, India and Terai region of Nepal

FOREWORD

The present work entitled 'Fluvial Geomorphology of Sapt Kosi River Basin' for which Sri Rajesh Kumar Mahato got his Ph.D. degree from Ranchi University, Ranchi in the faculty of Social Science is a good work on Fluvial Geomorphology. Thus, the outcome of a passionating research analysis following a gruelling task of field study, map making, diagram construction on the basis of compilation of data that sets a standard for similar- field based enquires at micro-level.

The Kosi has been a motorious river and has been called 'The sorrow of Bihar' for many years, for its flood havoc and frequent change of the channel. In order to check the frequency of flood havoc, the Govt. has almost finished the Kosi river project. But as in grosspart of the world floods do occur in plain areas.

The work has been divided into 12 chapters and the book has been suffixed by a bibliography which may have been called rich.

The technical aspect of the analysis of data and treatment of the subject is good; the photographs have been taken at the appropriate place to highlight the fluvial dynamics of the river which is in gradually not a single river but the combination of seven rivers, hence, Sapt Kosi river. The work drives a special mention for the geomorphic rendering the morphological related parameters for a visual depiction. The maps are well supported by tables and filed data.

The author could have added the references of the geomorphological works of Prof. Enayet Ahmad and Prof. D.P. Sathpathy who have devoted their whole service period in the research of geomorphology. Moreover the author did not seem to have consulted Geographical Outlook, a research Journal of the Geography department by Ranchi University whose a large number of papers on geomorphology have been published by the authors of this country and abroad.

Similar studies on the Fluvial Geomorphology on river basins of Bihar will aid in enhancing the geomorphological studies of depositional plain which up till now was mostly controlled on the erosional topography of the country.

21st September, 2017

D.P. Sathpathy
Professor & Former HOD,
University Dept. of Geography,
Ranchi University,
Ranchi, Jharkhand, India

PREFACE

Almost every dimension of human life is influenced by river system and their flowing water resource. Therefore, considerable interest has been evinced in the human society to understand the various aspects of river system such as their formation and fluvial system, geomorphic processes, drainage development, channel morphology and dynamics, hydrological status, sedimentation and bank erosion.

In order to understand the fact of fluvial geomorphology of Sapt Kosi river basin, basic function system and their applied aspect have been explained and analysed in this book. Sapt Kosi river system is one of the complex network in the world which is passing across rough terrain to rough climate system and provide large quantity of water for Gangatic river system. Hence, the text covers fluvial system of the river, their processing system, drainage development, morphometric analysis, channel morphology, channel dynamics, fluvial dynamics, geo-hydrological status, sedimentation, bank erosion and applied geomorphology on the river system. The text can be used for teaching course on fluvial geomorphology to graduate and undergraduate students in Geography and Geology and it can be used as reference material for allied sciences.

This book is spread over the 12 chapters which seek to elucidate purpose and method of present work. First Chapter is dealt with the basic concept of Fluvial Geomorphology. Second chapter of the book has been devoted to the Geographical description of the Sapt Kosi river basin. Chapter three deals with the geomorphic processes operate in the river basin whereas in Chapter four deals to drainage development in the basin. Chapter five is devoted to the detail morphometric analysis whereas in Chapter six of the book channel morphology has been dealt. In Chapter seven channel dynamics has been discussed whereas in Chapter eight details of fluvial dynamics has been dealt. Chapter nine is devoted to Geo-hydrological status whereas in Chapter ten sedimentation has been dealt. In Chapter eleven bank erosion has been dealt whereas in Chapter twelve Applied Geomorphology has been evaluated in detail.

Rajesh Kumar Mahato

Dr. Rajesh Kumar Mahato

ACKNOWLEDGMENT

It goes without saying that a work of this magnitude is never the work of single individual, but rather the collective effort of many for this region there are numbers of groups and individuals that must acknowledge for their assistance with this understanding. The river name 'KOSI or SAPT KOSI' should be understand as same river system. In many places (texts) are being used by both names. But in view of Kosi river basin and Sapt Kosi river basin, it is quite different, which is mentioned in chapter no. 2. I would like to thanks Dr. Jitendra Shukla, my Ph.D. mentor who guided me all through out the way, and responsible for my philosophical foundation on fluvial geomorphology of the concern region. I also want to like thanks to Dr. Ram Kumar Tiwari, Associate Professor, H.O.D., University Deptt. of Geography, R.U., Ranchi and Dr. Saroj Kumar Singh, Associate Prof., H.O.D., Deptt. of Geography, V.B.U., Hazaribagh for his valuable suggestions.

Additionally, I would like to acknowledge with deep sense of gratitude to all the agencies officials of the Nepal and Bihar and the concerned agencies involved in different purposes in the catchment area. In this connection firstly thank to Sri Rajesh Kumar, Chief engineer, Central Water Commission, Govt. of India, Supaul who provided full cooperation and assistance during the field work and has provided relevant materials (data and figures). Thanks are also goes to the officials of Nepal, DHM (Department of Hydrology and Meteorology) for providing relevant data and information. Other officials of offices like NATMO, Survey of India, Geological Survey of India, Block offices of north Bihar particularly Darbhanga, Madhubani, Supaul, Saharsa, Water resource and irrigation department, Govt. of Bihar along with countless others.

My deep sense of gratitude also goes to a numbers of foreign friends who assisted me during the field work like Parvat Iar (Nepal), Shankar Panthi (RS, Xishuangbanna tropical botanical garden, China) and Biswajit Nath (Assistant Prof. C.U., Bangladesh).

I am highly grateful to the IPCC (Intergovernmental Panel on Climate Change), Geneva and Foundation Cuomo, Monte Carlo for providing a project on 'Assessment of climate change over water resource of Sapt Kosi river basin', and Fellowship with full financial support.

I must thank to Mr. Sanjay Gupta of Rajesh Publications, New Delhi for efficiently bearing the responsibility for the publication of the book.

Lastly, thanks go to all those persons to whom I have forgotten to thanks them who has directly or indirectly helped me in the completion of this book.

Vijaydashmi
30th Sep., 2017

Dr. Rajesh Kumar Mahato
Asansol, West Bengal, India

ABBREVIATIONS

AFC	Altimetric Frequency Curve
AHP	Analytical Hierarchy Process
Am.	American
Amsl	Above mean sea level
APN	Asia Pacific Network
Asso.	Association
BIS	Bureau of Indian Standard
CBS	Central Bureau of Statistics (Nepal)
CMS	Consolidated Management Services (Nepal)
Co.	Company
Corp.	Corporation
CWC	Central Water Commission
DEM	Digital Elevation Model
DHM	Department of Hydrology and Meteorology
Edt.	Edited
Edn.	Edition
ENSO	El Nino-Southern Oscillation
FAO	The Food and Agriculture Organisation
FGI	Flow Geometry Index
Geog.	Geography
Geol.	Geology
Govt.	Government
GSI	Geological Survey of India
GWP	Global Water Partnership
Himl.	Himalaya
HSI	Hydraulic Sinuous Index

i.e.	Id est (That is)
Ibid.	Ibidem (in the same place)
ICIMOD	International Centre for Integrated Mountain Development
IGIS	Integrated Geographical Information System
IMI	International Meteorological Institute
IPCC	Intergovernmental Panel on Climate Change
IRD	Investigation and Research Division
IWMI	International Water Management Institute
MOPE	Ministry of Population and Environment
NAGI	National Association of Geographers India
NGO	Non Governmental Organisation
NHD	National Hydrography Dataset
NPHC	National Population and Housing Census (Nepal)
RMC	Regional Meteorological Centre
SOI	Survey of India
SSI	Standard Sinuosity Index
TAR	Tibet Autonomous Region
TM	Thematic Mapper (Landset)
TSI	Topographic Sinuous Index
UN	United Nations
UNDP	United Nations Development Program
UNEP	United Nations Environment Programme
WHO	World Health Organisation

CONTENTS

LIST OF FIGURES

LIST OF TABLES

CHAPTER – 1

FLUVIAL GEOMORPHOLOGY

The term Geomorphology has the Greek origin (Geo-Earth, Morpho-Form, and Logos-Knowledge) meaning the knowledge pertaining to landforms. The subject matter of Geomorphology includes land sculpturing processes and typology of landforms especially of the ground surface rather than the subsurface Geommorphology, therefore is defined as the science of description of various forms of the earth surface. To be more precise, forms in topographic features or geometric features of the earth surface. 'Geomorphology as the interpretative description of the relief feature of the earth surface (Worcester, P.G., 1948 [2nd Edn.]).

Landscapes are the product of complicated physiographic environment 'comprising cause-and-effect chain system' the geomorphic processes define the dynamic action or events in geomorphological system which involves in the application over face gradient. Thus, 'Geomorphology, stress the importance of the inter-actions at contact surface between the terrestrial parts of lithosphere and the liquid and the gaseous envelopes which surrounded it. It deals with over changing inter adjustments between different forces of the nature and the changes in the land surface as modify due to the dynamic balance between these forces (Morisawa, M.E., 1977).

'The process-oriented geomorphology is closely related to climatology, because air temperature, precipitation, winds and atmospheric humidity largely determine the response to the rocks to the subarial exposure (Bloom, A.L., 1978).'

Such a complex discipline, a geomorphology is has given rise to different stream of the branch.

i) **Climato-genetic Geomorphology:** An approach of

Geomorphology that emphasis the shaping of landforms in response to various climatic conditions (Birot, 1968; Tricart & Cailleux, 1972).

ii) Tectono-genetic Geomorphology: Branch of Geomorphology where emphasis is given to geology, structure and the tectonic movements in the evolution in the landforms.

iii) Quantitative Geomorphology: Where emphasis is given to explain landforms characteristics of such layer areas where sufficient evidences for historical study are not available because of destruction of relief forms due to greater degree of dissection of by subsequent geomorphological processes.

Geomorphology may deal with specific geomorphic processes (Leopold, 1964; Ollier, 1969; Birkland, 1974 and others).

Geomorphology is mainly said to...

i) Involve the interpretative description of the relief feature of the earth – Worcester, 1965.

ii) Deals with the surface of the earth's crust – Lobeck, 1939.

iii) With the surface of the earth surface feature their characteristics, origin and evolution – Challinor, 1969.

iv) Study of the origin and evolution of the relief feature of a landscape or landform visible upon the earth surface – Strahler, 1964.

v) Study of the features of the surface involving interpretative description of landforms their origin and development and nature and mechanism of geomorphic processes which involve the landform with a view that all landforms related to the particular geologic processes, or set of processes, and that the landform, thus developed may involve with time through a sequence of forms dependent on part, on the relative time a particular processes has been operating – Easterrook, 1969.

vi) Be the systematic description and analysis of landscape and the process that change them – Bloom, 1979, p. 5.

vii) Geomorphology involves in the study of a) Landforms and underlying rocks, b) the evolution of landscape, and c) actual processes (mainly erosive) which has produced the landforms (B.W. Sparkes, 1972).

Thus, the main aim of Geomorphology is to explain 'how and when' as well as 'what and where'. Only the alternate question of why is left to the philosopher. The ultimate goal of such study is a far more detailed understanding of morphological parameter of a drainage network such an understanding will be of interest for its own sake and will also be of assistance to the student of the hydrology and geohydrologists.

A river carries water, sediment and solid form the drainage area to the sea and thus the great interest to the geomorphologists, sedimentologists and hydraulic engineers. This is important to geomeorphologists because water is used by variety of purpose by humanity; water courses are used as navigation channels and also erosion, transportation and deposition of sediments cause a number of problem in the river and the catchment and that must be solve pragmatically.

Suspended sediment carries by flows is erosioned one of the significant measures of geomorphic effectiveness. Sapt Kosi river has remained a course to the mankind particularly in the lower reaches of Bihar plain. It has damaged the humanity from the beginning. Principal characteristics of suspended sediment concentration, suspended sediment load, and sediment yield during the monsoon region in Sapt Kosi basin have been a general mysteries and problems for the geomorphologists.

DEVELOPMENT OF GEOHYDRODYNAMICS: Even though mankind has been living with sediment problems for the past several centuries, relatively little progress was made in our knowledge about sediment movement up to 16th century A.D. Earlier civilizations in the valleys of the Indus, the Tigris, the Euphrates, the Nile and the Yellow rivers were using canals for supplying water for irrigation through unlined channels. These canals either took off from a weir or they were inundation canals. The common problem with these canals was silting and hence frequent sediment removal was necessary. Locating the canals on the outer side of the bend of a stream to reduce sediment entry into canal seems to have been practiced. The Chinese had made considerable progress in controlling large rivers, flood diversion, and similar other problems. The Romans had made progress in water supply and sewerage. The Greeks knew about the fall velocity of different sediment particles.

"During 1600-1800 A.D. relatively more progress was made in understanding the physics of flow in open channels. The basic equations governing the flow, viz. the continuity equation and the equations of motion were developed during this period. *d'* Alembert (1717-1783) gave the differential equation for continuity of flow which was generalization by Leonard Euler (1707-1783). It was also during this period that the equations of motion, commonly known as Euler's equations were established. Some basic ideas about river hydraulics were initiated by Dominico Guglielmini (1655-1710) and Paul Frizi; both wrote books on rivers. Du Buat (1734-1809) gave scouring velocity for materials of different sizes (Garde, R.J., 1995)."

Much more progress was made during 19th century. Bouniceau, Grass, Lechalas, Suchier and Deacon conducted studies and critical velocities for different sized materials were recommended. D.F. du Boys (1847-1924) gave a simple model for bed-load transport and reached the conclusion that $q_B \sim \hat{o}0(\hat{o}_0 - \hat{o}_{0c})$ were q_B is the rate of bed-load transport, and $\hat{o}_0$ and $\hat{o}_{0c}$ are the average bed shear stress and critical shear stress for given size of bed material respectively. In the later half of 19th century O. Fargue (1827-1910) who was closely associated with the developmental work of the river Garonne, gave what are popularly known as "Fargue's rule" of river behavior. It was also at the fag end of 19th century that Dennedy proposed the method for design of stable channels based on canal data from India that was later modified by Lacey and others.

The first half of the 20th century witnessed all round progress in fluvial hydraulics. G.K. Gilbert (Gilbert, G.K., 1909) performed extensive laboratory experiments and studied modes of sediment transport, and observed various bed-forms. Different investigators later used the hydraulic data collected by Gilbert to study resistance and sediment transport in channels. The credit for developing the rational criterion for incipient motion that is based on sound principles of fluid mechanics goes to A.F. Shields (Shields, A.F.,1965). In 1948 E. Meyer-Peter and R. Muller proposed an empirical equation for bed-load transport which is based on a wide range of sediment sizes and flow conditions and which is used often even today. A. Kalinske and H.A. Einstein (Einstein H.A., 1978) developed bed-load equations using statistical nature of sediment movement. Bed-load and suspended load samplers were developed and tested in Europe and U.S.A., which greatly helped in collecting valuable data on sediment transport by rivers.

As regards the resistance to flow, Karman and Prandtl's equations for velocity distribution for turbulent flow in pipes were adapted to open channel flow and velocity distribution laws for hydrodynamically smooth and rough surfaces were established. Einstein (1904-1963) suggested a method for separating grain resistance and form resistance of bed undulations while Einsten and Barbarossa proposed a method for predicting resistance to flow in alluvial streams.

Lastly on the basis of a large volume of data from stable mobile bed alluvial channels and building on the advances made by Kennedy, Lindley, King and others, G. Lacey (1887-1980) proposed a method of channel design according to which for given Q and bed material size, the channel depth, width and slope are uniquely fixed. Also data were collected about the geometry of alluvial rivers and equations have been developed to predict width and depth as a fuction of badkful discharge and sediment size.

Kennedy, Engelund, Hansen and Fredsoe, have studied stability of mobile bed subjected to small disturbances to explain the formation of dunes, antidunes and plane bed. Similarly, Hansen, Callander, Parker, Hayaski and Ozaki, Engelund and Skovgaard and others have carried out stability analysis to determine the conditions under which streams meander.

And finaly, field data are being collected to test various softwares developed for solving such problems. Also experimental data are being collected to study some basic problems such as armoring and pickup function.

HISTORICAL DEVELOPMENT OF FLUVIAL GEOMORPHOLOGY (Tinkler, K.J., 1985): one can extract three basic principles regarding the rational investigations of landforms; these are i) the concept of infinite time, ii) reality of denudation, iii) acceptance of the principle of conservation of mass. Herodotus (485-425 B.C.) recognized the importance of yearly increments of silt and clay deposition by the Nile. He also anticipated the idea of changing sea levels that is of great significance in geomorphology. Aristotle (384-322 B.C.) thought that rainfall might produce a temporary torrent, but doubted that it could maintain flow in a river. Strabo (54 B.C. – 25 A.D.) noted examples of local sinking and rise of the land. He also mentioned about the effect of ebbs and tides on the growth of delta. Both Strabo and Seneca (B.C. – 65 A.D.) recognized the role played by volcanic activity and earthquakes on the landforms.

During the many centuries that followed the decline of the Roman Empire, there was little or no progress of scientific thought in Europe; however, some learning process continued in Arabia. During 941-982 A.D. there is reference to erosion and transportation of sediment by the streams and wind and weathering in the four-volume tretise on discourses of the Brothers of Purity. Little progress was made in Europe between the first century and beginning of the 16th century. During fifteenth, sixteenth centuries landforms were explained by the philosophy of catastrophism, according to which the features of the earth were created as a result of violent catastrophic actions. Leonardo da Vinci (1452-1519 A.D.) had very advanced ideas about geologic thinking for his time. He recognized that streams cut the valleys and that the streams carried sediment from one part of the earth and deposited at other places. The Frenchman Baffon (1707-1788) thought that erosion by streams would eventually reduce the land to the sea level. He was also the first to suggest that the age of the earth was not to be measured in terms of a few thousand years. Another Frenchman Guetthard (1715-1786) also discussed about the degradation of mountains by streams and emphasized that not all the material removed by the stream would immediately be carried to the sea but a part would also deposit on the flood plains. The Swiss De Saussure (1740-1799) recognized the ability of glaciers to carry out erosional works.

James Hutton (1726-1797) who is known for propounding the concept that "the present is the key to the past", thus establishing the doctrine of uniformitarianism. His writings clearly express the concept of a river system and its geomorphic significance. Some other important concepts introduced by Hutton are:

i) A vast portion of the present rocks is composed of bodies, animals, vegetables and minerals of more ancient formation.

ii) All present rocks are going to decay and their material going to deposit in the sea.

iii) The morphological process requires indefinitely long geological time.

iv) There is a conceptual possibility of relative change between land and sea levels leading to upheaval.

Hutton's friend John Playfair (1748-1819) published Hutton's work namely 'Illustrations of Huttonian Theory of Earth' in 1802. Playfair proclaimed the ability of glaciers to erode their valleys deeply.

Sir Charles Lyell (1797-1875) was somewhat doubtful about the immense ability of running water to carve the valleys. It was during 19th century that there was recognition of an ice age during which much of North Europe was covered with ice sheets. Playfair had sensed the possibility of large boulders being transported by glaciers. Louis Agassiz (1807-1873), Venetz of Switzerland in 1821, Bernardi of Germany in 1832 and Jean de Charpentier in 1836 supported this concept of glaciations in Europe. In the later part of 19th century books were written to describe the principles of landform development. These were by Peschel, Richthofen and A. Penck.

The basic foundation of geomorphology was laid by America in later half of 19th century by Major J. W. Powell (1834-1902), G.K. Gilbert (1843-1918)[5] and C.E. Dutton (1841-1912). Powell's studies of Unita Mountains emphasized the importance of geologic structure in the classification of landforms. He also introduced the concept of the limiting level to which the land-level would reduce and called it the base level. Col. George Greenwood earlier used this concept in Europe in 1857. Powell recognized that the process of erosion, if carried undisturbed on land, would reduce it eventually to a level little above sea level. He was to correctly interpret that various unconformities in rocks in the Grand Canyon, Colorado (U.S.A.) correspond to ancient periods of land erosion.

G.K. Gilbert's contribution in experimental work carried out in California has already been described. He was a pioneer in studying hydraulic mining and its effect on stream morphology. His other contributions includes recognizing the importance of lateral planation by streams in the development of valleys and his explanation of Henry Mountains of Utah (U.S.A.) as the result of erosion of intrusive bodies. Dutton gave a penetrating analysis of individual landforms. Gilbert and Dutton are given credit for initiating the concept of erosional unloading of the earth's crest technically known as isostasy. W.M. Davis (1850-1934) had greater impact on the development of geomorphology than anyone else. Of all the contributions to geomorphology, Davis is remembered for introducing the concept of geomorphic cycle. According to this concept in the evolution of landscapes there is a systematic sequence that enables one to recognize the stages of development of landforms. The sequence is called by him as youth, mature and old age. These landforms are explainable in terms of differences in geologic structure, geomorphic processes and the stage of development. In the development of the idea of geomorphic cycle Davis had assumed that there is a relatively rapid uplift due to diastrophism which is followed by a relatively long period of standstill which permits the erosion cycle to run its course. W. Penck and his followers questioned Davis's idea of geomorphic cycle during 1920's and 30's. In spite of these objections the Davisian geomorphic cycle is still considered a reasonable model primarily because of the absence of a plausible reasonable alternative.

RECENT CONTRIBUTIONS: Since the end of the Second World War a large number of aspects about river morphology and hydrology have been or are being studied. These include channel geometry, mathamtical modeling, effect of neo-tectonics and mass movements on channels, fluvial systems, experimental fluvial morphology, palaeo climatic and palaeo hydrologic effects and gravel-bed rivers. Scientists working at U.S. Geological Survey have studied short-term morphology of river channels; they include W.B. Langbein, L.B. Leopold and M.G. Wolman. S.A. Schumm, M.P. Mosley and W.E. Weaver studied fluvial systems and performed experiments in the laboratory to study river morphology. J.R.L. Allen from U.K. has done extensive work on the character and classification of bed forms and sedimentary structures with respect to deltas, meanders and floodplains. Many investigators including K.J. Gregory, J. Lewin, V.R. Baker and L. Starkel have studied Palaeo climatic and palaeo hydrologic effects on river channels. Geographers in U.K. have given impetus to the research in gravel-bed rivers and this work is now continued in Canada, U.S.A. and New Zealand.

HISTORICAL DEVELOPMENT OF FLUVIAL GEOMORPHOLOGY IN INDIA: Fluvial geomorphology in real sense, involves the study of river (fluvial) dynamics (including runoff, discharge and erosion etc.), transport of suspended sediment load and bed-load, mechanism of sedimentation , river morphosis, channel forms and pattern, resultant landforms etc. on the basis of primary data derived through field measurement and mapping, instrumentation and laboratory experiments and analysis. K.R. Dikshit, on this parameter, has rightly pointed out that 'fluvial geomorphology has not received the attention it deserves from Indian geomorphologists. In fact, there did not simply exist a tradition of fluvial geomeorpholgy in India till Leopold's fluvial processes in geomorphology' (book) appeared on the scene and gave a new orientation to the discipline' (K.R. Dikshit, 1983, p. 5). Most of the works on fluvial geomorphology, rather river geomorphology, are related to the evolution and development of drainage and drainage pattern, calculation and depiction of longitudinal and transverse profiles on the basis of topographical maps, calculation of sinuosity indices and meander properties and quantitative description of fluvially originated landforms. Recently, field measurement and instrumentation of fluvial processes have been initiated at Allahabad (Savindra Singh and Alok Dubey, 1991-95), Pune (V.S. Kale), Gauhati (D. C. Goswami), Almora (J.S. Rawat) etc. and micro-level studies of fluvial system have been reported.

Significant contributions in evolution and development of river basins and drainage pattern have been made by H.L. Chibber, (1954: drainage pattern observed in India and adjacent countries, 1954: thrust superimposed drainage of the Shali area), K. Bagchi, (1960: drainage pattern in Ganga delta), B. Ghose and S. Singh, (1963: reconstructed drainage system in central Luni basin), A.B. Mukherjee, (1970: terminal-trangular tract of inland streams in Haryana), A.K. Pal (1970: analysis of Balsan river basin, West Bengal), S.R. Basu and S.C. Chakraborty (1972: some consideration on the decay of the Bhagirathi drainage system), K.M. Hironi (1982: evolution of the antecedent drainage pattern: a study of Banas), B.P. Radhakrishna (1983: evolution of the Sharwati drainage, Mysore State), Amal kar and B. Ghose (1984: the Drishadavati river system of India: assessment and new finding), R.M. Pofali (1970: pedogeomorphic problems of the Purana basin, Maharashtra), A.B. Mukherjee (1993: morphology and drainage evolution of Saraswati basin) etc.

The forms and characteristics of drainage basins of Konkan have been thoroughly examined by K.R. Dikshit, 1976; S.R. Basu and L. Ghatowar, 1986; V.S. Kale and S.N. Rajguru, 1989; S.R. Basu, 1967; S.R. Basu and N.R. Kar, 1970; N. Prasad, 1992; S. Singh, 1984; S.P. Chatterjee, 1947; and B.S. Marh, 1989.

Fluvial system and fluvial dynamics of some rivers have been thoroughly examined by S.R. Basu (1967), while S.R. Basu and L. Ghatowar (1986) studied long profiles of the Lish and Gish Rivers of Darjeeling Himalaya, V.S. Kale and S. N. Rajguru (1989), have presented a critical analysis of longitudinal stream profiles of western upland, Maharashtra and its implications on landscape stability. River meandering has been studied by S.R. Basu (1967: fundamental problems of meander formation with special reference to Bhagirathi), S.M.K. S.R. Basu and N.R. Kar (1970: meandering, migration and alluvial terraces in lower Bhagirathi valley), N. Prasad (1992: some aspects of meandering streams of the Barakar basin and their sinuosity indices), S. Singh (1984: dynamics and morphology of meandering rivers in western Rajasthan), Changes in river courses and problems arising therefrom have been documented by S.P. Chatterjee (1947: river problems in Bengal), H.L. Chibber (1950: westerly drift rivers of northern India), S.R. Basu (systematic study of the changes of the river Bhagirathi-Hugli and its impact on the port of Calcutta) while B.S. Marh (1989) has described the river capture in the Himalaya..

Fluvial system and fluvial dynamics of some rivers have been studied by Indian geomorpholgists mostly based on secondary data e.g. S.R. Basu (1967: some aspects of fluvial dyanamics of river Bhagirathi valley), M. K. Bandopadhyay (1982: river dynamics of the Ganga delta), S.R. Basu, (1984: study of fluvial dynamics of the Bhagirathi river), S.N. Rajguru and V.S. Kale (1985: changes in the fluvial regime of Western Maharashra upland rivers during late quaternary), R.K. Pande (1986: evaluation of Gole river annual discharge: an experience of springfed Siwalik mountain river), D.C. Goswami (1989: magnitude and frequency of fluvial process in the Brahmputra basin), S.C. Mukhopadhyay (1993: fluvial environment of Mechi-Sankosh interfluves), and S.N. Rajguru, V.S. Kale and G.L. Badam (1993: quaternary fluvial systems in upland Maharashtra) etc.

Fluvial erosion by streams and gullies has been scantly studied by Indian geomorphologists but some contributions have been made by K.B. Ray (1954: erosion of river banks: its causes and prevention), V. Subramaniam and R. Dalvi (1978: denudation by Indian rivers; 1978: some aspects of stream erosion in Himalayas), M.K. Bandopadhyay and N.A. Singh (1984: erosional characteristics of the Manipur river basin), M.K. Bandopadhyay (1984: measurement of erosion under fluvial processes in some selected areas of eastern India), Savindra Singh and A. Dubey (1988: fluvial erosion and sediment load characteristics of Ganga system in U.P.), S. Rawat and M.S. Rawat (1984: accelerated erosion and denudation in the Nana Kosi watershed, central Himalaya), Savindra Singh and Neera Rastogi (1994: study of morphodynamics and morphogenesis in sub-humid tropical environment) etc.

Channel pattern, Channel form and Channel morphology have been studied by V.S. Kale, S.N. Karlekar and L.A. Deodhar (1986: channel morphology and hydraulic characteristics of Vashishthi river, Maharashtra), V.S. Kale and S.N. Rajguru (1988: effect of channel perimeter lithology on the morphology of upland Maharashtra rivers), J.S. Rawat (1988: anthropogenic transformation of channel network capacity: an experimental study in Kumaun Himalaya), V.S.Kale and S.N. Karlekar (1989: relief and drainage network morphology of Mandvi and Vashishthi river basins), R.K. Pandey and D.S. Rawat (1989: channel development in quartzitic limestone of Rameshwar formation-central Himalaya), V.S.Kale (1990: morphological and hydrological characteristics of some allochthonous river channels, western Deccan trap upland region), S.N. Rajguru et. al (1995: channel form and processes of flood dominated Narmada river), V.S.Kale et.al. (1996: multi-channel pattern of bedrock rivers: an examples from central Narmada basin) etc.

In the field of alluvial morphology, sedimentation and depositional fluvial landforms A.B. Mukherjee (1964) is of paramount significance as he has elaborated the alluvial morphological characteristics of the upper Ganga-Yamuna doab. Alluvial and sub-alluvial morphology of the Western upland rivers of Maharashtra has been analysed by V.S. Kale and S.N. Rajguru (1985) who have also traced (1987) late quaternary alluvial history of the north-western Deccan upland region. Kale's study of colluviums in the upper Krishna basin has acclaimed popularity. P.K. Sen (1989) has studied the alluvial morphology of the lower Damodar basin.

A few geomorphologists have studied different aspects of alluvial fans eg. Alluvial fan model for Himalayan piedmont deposits by S.S. Roy, 1981, alluvial fans in the northern Brahmputra valley by L.H. Dutt, 1984; the Chandigarh Dun alluvial fans: an analysis of form-processes relationship, by A.B. Mukherjee (1989), a very exhaustive and analytical study of alluvial fans, 'development of alluvial fans in the foothills of the Darjeeling Himalayas and their geomorphological characteristics' by S.R.Basu (1990) etc. Sedimentation process and some micro-depositional landforms have been described by S.R. Basu (1972): on the formation of a shoal on the concave bank of the laterite river Kopai, Shantiniketan, S.R. Basu (1981): process of sedimentation in the Hooghly tidal channel, B. Ghose et. al, 1983, coparative role of the Aravalli and the Himalayan river systems in the fluvial sedimentation of the Rajasthan desert, V.S. Kale, 1980, channel fill deposits of upper Godawari basin, S.N. Rajguru and V.S. Kale, 1991, quaternary fluvial sedimentation in the semi-arid rivers of upland Maharashtra, S.R. Basu, 1967, quantitative analysis of silt behavior of the river Bhagirathi, etc.

Data morphology has been described by M.S.Rao and R. Vaidyanadhan (1979: morphology and evolution of Godawari delta), M.S. Rao, K.N. Rao and R. Vaidyanandhan (1978: morphology and evolution of Mahanadi and Brahmani-Baitarni deltas), M.S. Rao (1982: morphology and evolution of northern Cauvery delta), J. Seetharamaiah and A.S. Swami (1989: sedimentatological characteristics of landforms in Pennar delta) etc. Flood morphology and flood dynamics have been studied by B. Ghose and S. Singh (1965: geomorphic changes in flood plains of Luni basin), B.C. Acharya (1970: Mahanadi and its floods), Savindra Singh (1983: flood hazards and environmental degradation), S.M. Bhamre (1989: Arni flood terrace: a sedimentological analysis), V.S. Kale (1990: flood geomorphology of allochthonous river channels of western upland Maharashtra), V.S. Kale et. al (1993: flood geomorphology of Indian Peninsualr rivers), V.S. Kale et. al (1994: geomorphic and hydrologic aspects of monsoon flood on the Narmada and Tapi rivers in central India). River terraces have been widely studied by Indian geomorphologists but a few research papers have appered e.g. 'river terraces of Yamuna and ton' by D.Niyogi (1970), 'river terraces of the Subarnrekha basin' by S.C. Mukhopadhya (1973), 'Choe terraces of the Chandigarh Siwalik hills by A.B. Mukherjee' etc. 'Rejuvenation and polycyclic fluvial landforms have been studied by J.S. Rawat et. al (1984), R.K. Rai (1986), G. Padmja (1994, while palaeo-drainage and channels have been studied by R.P. Verma and P.K. Roy, Luni river basin) and V.N. Bajpai and H. Kumar (1988: aquifer disposition and palaeo-channel morphology in Kanpur-Unnao region of Central Ganga plain).

It is significant to point out that up to the year 1970 elementary studies of alluvial monsoon rivers were carried out by the earth scientists while significant contributions came from hydraulic engineers. In fact, the study of fluvial geomorphology of Indian rivers began after 1970, to be more specific after 1980. Mukherjee (1976), Goswami (1985), Gupta and Dutt (1989), Kale

(1990), Sinha and Friend (1994), Gupta (1995), Savindra Singh and S.P. Agnihotri (1987), Savindra Singh and A. Dubey (1988), Savindra Singh and A. Dubey (1991), Savindra Singh and A. Dubey (1996), Savindra Singh and A. Dubey (1997a) and (1997b), Savindra Singh and A. Dubey (2000), Kale and Hire (2004), Sinha et. al. (2005), Sharma (2005) etc have made significant contributions in fluvial geomorphology. Besides systematic studies of major Indian rivers, new fields such as fluvial dynamics, quaternary fluvial geomorphology, bedrock rivers, palaeo-hydrology, flood geomorphology, channel geomorphology and patterns, fllod hydraulics and dynamics, flow hydraulics and sediment transport etc. have emerged in fluvial geomorphology.

Enormous works have been done in different fields of fluvial geomorphology by Indian geomorhpologists but it is not practically possible to enumerate all these works. A critical review of the status of fluvial geomoprhoogy of Indian rivers has been carried out by V.S. Kale (2002). In the first decade of the 21st century significant contributions in different fields of geomorrpholgoy have come from Harijan et. al. (2003), Jain and Tondon (2003), Jain and Sinha (2004), V.S. Kale (2003), Kale and Baraker (2006), Kale and Joshi (2004), Kale and Hire (2004), V.S. Kale et. al. (2000), V.S. Kale (2003), B.M. Maurya et. al (2003), S. Misra et. al. (2003), Sharma and Phukan (2004), Sarmah (2003), J.K. Tripathi et.al (2004), K.S. Valdia (2002), R.J. Wasson (2003), M.Jade et.at. (2005) etc.

GEOMORPHIC STUDIES AT THE REGIONAL LEVEL: 'Regional Geomorphology, like regional geography, is by far the most common topic by Indian geomorphologists. Most geomorphologists start with the study of regional details and subsequently branch off to an analytical study of any specific area of the discipline. Those more geomorphological studies are concerned with landscape evolution in different regions (K.R. Dikshit, 1993, p.17). The Indian geomorphologists found peninsular India in general and Chota Nagpur region in particular as geomorphological playgrounds for geomorphological investigations. R.P. Singh studied different geomorphological aspects of Chota Nagpur in much detail and varied geomorphic features thereof. Thus, he is given credit for presenting well documented and comprehensive accounts of geomorphology of Chota Nagpur. Sarvindra Singh, having a definite and well organized research design, has attempted to present detailed geomorphological accounts of systematic, regional, quantitative, theoretical and applied geomorphology of the foreland area of Indian peninsula. Keeping this fact in mind he divided the foreland area into smaller physiographic regions and assigned each region to his research student's doctoral work. This scheme resulted into the presentation of well documented regional geomorphologyical accounts in the form of Ph.D. thesis and research papers resulting therefrom covering the foreland area from S.E. Chotanagpur region in the east to Aravallis in the west.

A lot of research work has been done recently in the department of Geography, R.U., Ranchi in the ralm of fluvial geomorphology. Mention may be headere for the research work of Jitendra Shukla, (1996: Geomorphology of Palamu uplands); M.P. Prajapati (1997: Geomorphology of Karkari basin).

Mentioned may be may for the work of development of landform in the Himalaya, 1951 by Dr. H.L. Chibber; S.L. Kasyatha, 1955: Development of landforms and courses in Shivalik range; D. Niyogi, 1957: The Himalaya Beaj basin-A hydrological study; G.K. Dutt, 1982: Landforms of lower Ganga basin; S.C. Mukhopadhyay, 1991: the complex structure and geomorphological analysis of central Himalaya and surroundings; S.R. Basu and M. Kuahle, 1992: Geomorphology of east Karakoram, west Himalaya etc.

CHAPTER – 2

SAPT KOSI RIVER BASIN

Sapt Kosi river basin an important tributary of Ganga river system form by the confluence of the three river i.e. Sunkosi, Tamur, and Arun, which originates in the snow covered peaks of Sikkim, Nepal and Tibet. They are fed by glacier of Gauri Shankar-Everest-Makalu-Kanchenzonga group of peaks. These four peaks contributes respectively 43%, 44%, 37% and 19% of the total water flow in the river, all these three river unite north of the Mahabharat range to form the Kosi. The river enters the Terai of Nepal after cutting a narrow gorge of 10 km in the Mahabharat range at Chatra.

EXTENSION: Sapt Kosi river which basically originates from the Tibet plateau flowing in easterly direction through the China at Khimathangka La enters Nepal and flows in southern direction. It flows 170 km in Nepal and enters India at Birpur

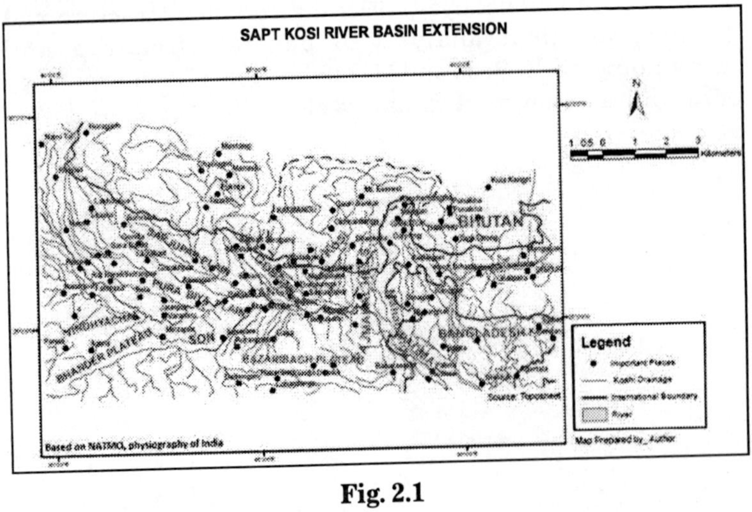

Fig. 2.1

(26.53'00"N & 86.25'00"E). The total drainage area of the basin 91,476 sq. km out of which 21,500 sq. km lies in India "Kosi is a major tributary of the Ganga river system and has long been consider as a problem river due to recurrent and extensive flooding and frequent changing in course. It has frequently changed its course in westerly direction and has latterly moved nearly 150 km.

Sapt Kosi river basin lies between $25^0$24'N to $30^0$27'n latitude and $84^0$4'E to $88^0$12'E longitude. It is roughly rectangular in shape and has a longitudinal extent of 885 km east west and 145 -241 km north to south. The basin is extended over the country of India, Nepal and China and physiographically covers the Tibet plateau, Greater Himalaya, the Middle Himalaya, Lesser Himalaya, the Shivalik and extensive flood plain of India. Sapt Kosi river basin is bordered by the Brahmputra (Psangpo) river in the north, the main stream of Ganga river in the south, Mahananda river and its basin area in the east and the Gandak river the tributary of the Ganga river in west. Sapt Kosi river is a trans-boundary river between China, Nepal and India signifies by one of the largest tributary of the Ganga. The river is joined by major tributary approximately north of the Indo-Nepal boarder breaking into more than 12 distinct channels with shifted course due to flooding by Kamla-Balan, Budi Gandak (Past tributary). Major tributaries of the river Sapt Kosi in India and several other small tributary over the last 250 years, Sapt Kosi river has shifted its course westward more than 120 km. the river is characterised by heavy siltation in the lower reaches which are mostly carries by the tributaries along with the main stream during the Monsoon season. Sapt Kosi along with its tributaries, drains in the plains of the Bihar and known as the 'Sorrow of Bihar' a severely flood prone area in India. The frequent change of its course has caused and wide spread human sufferings from the very beginning.

PHYSICAL SETTING:

GEOLOGY: Sapt Kosi river basin extended over a wide area covering India, Nepal and China. The basin is geologically very diverse and complex. Southern Tibet, Greater Himalaya, middle Himalaya, Shivalik and the flood prone area track of India may be said to be very younger to the oldest rocks of Archean group.

"The Archeans are here represented by Daling and Darjeeling series. The Daling series is a Schistose group; it grads through a transition zone into dominantly gneissic Darjeeling series (Krishanan, 2011)."

The basin area has a long spell of geological history. Some geologist are of the opinion 'the formation of great Himalaya was completed 70 million years ago the scared phase was placed about 25-30 million years ago when the Middle Himalaya formed, the Shivalik form into last phase of Himalayan orogenic say about 2-20 million years ago. 'The basin covers six geological belts and has the main strata developed from the quaternary system to the Proterozoic (Owen, 1995).' 'The southernmost Ganga plain is made of the quarterly deposit with a wide range of Molassie and Shivalik gravels accumulated in the south of the lower Himalaya, which are subject of erosion from flash flood and thin debris flows, the rich gravel layers cover the area of Sindhuli and Rajbas (Sinha, Tandon, Gibling, Bhattacharjee and Dasgupta, 2005). 'There are widely developed schists and slate rocks in the lower south of the Himalaya, which are source material of debris flow and landslide in this arid area (Chen, Hu, Deng, Khanal, Zhu and Han, 2013).' 'It has been reveal recently a rich deposit of slaotes in an area from Kathmandu to Khuhot in the north of middle Himalaya there are widely deposits of igneous and metamorphic rocks such as granite, amphibolites, gneisses, magnetite etc. in the Zangmu-Rylam area, along the left bank there is abounded granite gneiss (Liu, 2006).' 'In the middle of the high Himalaya, there are very hard rocks such as granite and granite gneiss. In the north there is mainly a set of sedimentary rocks deposits of slates and schist including sandstone and limestone. The Tibetan Himalayas has a large area of sedimentary rock and small area of metamorphic rocks. The area from Tingri to Mt. Everest base camp has may soft rocks layers of well develop schist and slate etc (Liu, 2006).' The regional structure from the north-west to south west has significant fault 'the main boundary fault divides thick layer of gravel deposits and the metamorphic schist slate in the south side. The main central fault promoted hanging wall of granite to become roof of the world (Bilham, Gaur and Molhar, 2001).' The rock of both side of the Yarlung Xango suture zone have crust against each other and suffered from severe weathering.

1. The three element of the great basement of Himalaya are the mix up and may be the but may describe as first the metamorphosed sedimentary archean,

2. Intrusive granite and gneisses of the later period,

3. Reminents of archean granites, granulites, arthogneisses and schists.

In Sapt Kosi river basin Triassic system is also developed sedimentary formation this system is best developed in inner Himalaya from Hazra to Nepal. It constitutes the great scarp plateau of Tibet. A perfect section of these rocks, showing the relatec of trace to the system below and above. It is exposed at Liang in Chinese part of this basin. 'Limestone are the perennial system of the composition. The rocks are of light blue and great tint compact and homogenous, and sometime dolomitic in composition (Hein & Gansser, 1939).' They composed a picture feature of the Himalayas, notable from all part of county, by the light collaboration of the out crops and their graceful long and allegation folds. Rocks of the Jurassic formation are also dominant in the numbinates track of the basin. The Jurassic rocks are found to occur as Catral series and Umia series. The basin is also having extensive deposits of the cretaceous rocks. Crutens limestone is overland by great developed 120 of well-bedded-gray limestone succeeded by limestone of Eocene system. Later on the rocks of the Shivalik and range have been formed sandstone with sales, clays and conghom rills are the dominant rocks, which is overlain by the rocks of the middle Shivalik. Three formation of the middle Shivalik viz Tawakhola, Biotite tourmaline granites and Ghan Pokhra formation which is mainly form of sandstone and conglomerate boitite and interbeded by quartizes. Below the Shivaliks and extensive plain track of Sapt Kosi river basin composed mainly of the fine alluvium in on the extreme lower reached and alluvial, with boulders, gravels, sands and clays in the Terai region.

EARTHQUAKE: Seismic activities in the basin are also frequent and the reason is classify in eighth degree zone (Bilham at. al. 2001). In the last two hundred year two major earthquake with magnitude greater than 8.0 happened to be experience. First one happened on the 26 August 1933 at Shisha Pengma, west of Ngalar (28^{0}30N' and 85^{0}50'E) with a seismic intensity of 10^0 and second one occur on 5th January 1934 at Dabangia between Nepal and India, moreover a magnitude of 7.0 earthquake occur on Pulary (32^{0}2'N and 81^{0}2'E) in October 1883 with seismic intensity of 90 and magnitude of 6.25 occur at Tingri. Earthquake and low intensity seismic activity has deeply affected the shifting of Kosi main flow channel towards west. Many geologists are of the opinion that the repeated occurrence of earthquakes are responsible for the westward shifting of the main channels as the main flow of Sapt Kosi river has shifted at about 200 km westwards. Several researchers have been done on the shifting of the river course in which earthquake is also a cause among them.

PHYSIOGRAPHY/RELIEF: Sapt Kosi river basin characterised by diverse and very complex topographic conditions, as the basin relief charactersed by high altitudinal mountainous region full of peaks and passes with a shape of 90^0 and sometimes more than 90^0 form south to north. The terrain generally ascends till Mt. Everest and then descends. The southern region with an altitude between 2,000 m is named as Small Himalaya and bed rock is known as Shivaliks the area from the north to Sapt Kosi river to Nyalam with the altitude between 2,000-3,000 m is named as Middle Himalaya. The area from north to Nyalam with altitude of 3,000 m is known as Greater Himalaya. The Tibetan Himalaya refers to the region north of Tingri.

The relief features of Sapt Kosi river basin particularly in the hill region of Nepal Himalaya is developed as a recent between tectonic rock type and climate. The area of the Sapt Kosi river basin covering the Nepal Himalaya rose the mountain region due to tectonic movement in four phases of upheaval from Paleocene to Pleistocene. This has given the part of Himalaya to be developed as are the concave side facing.

TABLE - 2.1
PHYSIOGRAPHIC DIVISIONS AND THEIR AREA-PERCENTAGE OF TOTAL AREA

S.N.	PHYSIOGRAPHIC DIVISION	AREA IN KM (SQ. KM)	PERCENTAGE OF AREA (%)
01	Tibetian Plateau	12,070	13
02	The High Himalaya	17,800	19
03	The Midland Hills	9,232	10
04	The Mahabharata	8,300	9
05	Lekh Range	5,000	5
06	The Churia	6,000	6
07	Shivalik	10,542	11
08	The Plain	22,532	24
TOTAL		**91,476**	**100**

Source: Computed from IGIS software (version 1.1) with the help of produced map.

On the basis of variations of topographic conditions the whole Sapt Kosi river basin may be divided into eight physiographic regions.

1. Tibetian Plateau, 2. The High Himalaya, 3. The Midland Hills 4. The Mahabharata, 5. Lekh Range, 6. The Churia, 7. Shivalik and 8. The Plain

The above physiographic division has a regional shape and high variation in terrain features. All the above physiographic divisions characterised by very complex terrain features which has been presented in the table no. 2.1.

1. Tibetian Plateau: The Tibetian plateau area is the uppermost part of Sapt Kosi river basin rise over the north of Great Himalaya range. It is the origin place of the Arun river. Geologically, the plateau area is a rigid mass, the special characteristics of this region is permafrost, the area is over lain by the fluctuating snow in the surface, several hummocs and peri-glacial feature are abundantly found. This part of physiographic region having altitude of more than 3,500 m. the range of altitude is highly variable in the different part.

2. **The Higher Himalaya:** The Higher Himalayan zone covers an extensive area of Nepal. The midland zone in the north slowly rises in the hill range which average altitude go beyond 3,300 m. the area is mostly snow covered, glacier are found at the altitude of 3,300 m. The ranges are not contires, but parallel in nature.

Most of the peaks and passes are above 5,000 m from MSL. This ranges in a belt of approximately 40 km wide in which topography ranges from 8,848 m Mt. Everest to the minimum of 5,000 m in the river valley. In this range most of rocks are gneiss, granite, magmatite with schist and phyllite. The peak of the Mt. Everest is mostly composed of dolomite and quartzite, the region is dominated by 'U' shaped valleys, avalanches as well as the glacier and peri-glaciers landforms in abundance. Snow line occurs at average 4,000 m. A deep gorge is found at 27⁰48'N and 87⁰28'E of the Mt. Everest. The highest peak region of Mt. Everest the rock are mostly biotite schist and hornblands, sidmenite gneiss, micagneiss, garnet mica gneiss and augan gneiss. Rocks go deep towards north and north east. most of the hight peak of the Himalaya lies in this physiographic zones, they are Mt. Everest (Sagarmatha) 8,848 m, Kanchanjunga (8,586 m), Makalu (8,463 m), Annapurna (8,163 m), lies along the northern border of river basin and extended over Tibetian plateau in far northern river peripheral zone.

3. **The Midland Hills:** North of the Mahabharat range and south of Higher Himalaya lies the midland part of the low lying hills, rivers and tectonic valleys. This belt is approximately of 50 km in width. Extending from east to the west Nepal. It is well developed in the Arun river, which is the main part of Sapt Kosi river basin.

It is well developed in the Sun Kosi valleys and fall in a straight line Bhojpur, Okhladunga, Ramechap all fall in the northern edge of the Sun Kosi valley and connecting Panchkhal of the Kathmandu tectonic valley. Kathmandu is 1,220 and Trisuli is 610 m from the MSL. Tectonically, midland is also very complex and well developed in 30 km as the valley is low lying in Nepal Himalaya is interesting among the tectonic valleys. It has having the altitude of 2,000 m. extending from east to west Nepal, the temperature and climate of this region appears to be vary favorite in Nepal. The important streams are Dailekh, Silkrahi and Doti etc.

4. **Mahabharata Range:** It's lying between Churia hill of south and low lying area of midland of the north. Mahabharat range is an important physiographic feature of Sapt Kosi river basin. It attains height to 3,000 Meter. from MSL. From this hill several interminant and advetisiours streams gaits its origin and esimilates, Arun the main flow of Sapt Kosi river basin. Mahabharat range is located in the central of the eastern Nepal. The important peaks of the Mahabharat range are Hanspokhari, Sanguri, Lekh; all are above 1,830 meter. The western Mahabharat range is not the part of the Sapt Kosi river basin. It is also not well developed. The Mahabharat range has a width of nearly 32-50 km in the east Nepal, which is completely under the Kosi river basin, Tamur and Sun Kosi are the main part of the basin. The Mahabharat range rise steeply in the southern slope where as northern slope is gradual. The northern limit of this is mostly a river valley, this range sharply rises the southern face of this range is steep and the terrain is very rugged. Bagmati and Kamla the important tributaries of Sapt Kosi river originate from this range. The Mahabharat range acts as water parting hydrologically, as well as meteorologically; it acts as an accelerator to move the cloudy winds to the western Nepal.

5. **Shivalik range (Churia hills):** Churia or shivalik hill range is an important physiographic unit of Sapt Kosi river basin lies north of Terai region and rises abruptly 1,220 meter and runs east to west Nepal nearly 800 km. It merged with Mahabharat range and partly separate out from the Kankai river basin. It makes a separate range and bifurcates two ranges inclosing in Dun type valley west of Kosi. Churia makes two ranges and encloses two 'Dun' type valleys that is i) Trijunga located in the west of Sapt Kosi, ii) Rapti valley lies in the north and just south of Dang.

The width of Churia hill is 8 km in the east. Churia river gets rise from this hill which carries a huge amount of sediments from the Churia hills owing to normally low height the area is full of shorea robusta (Sal) forest. The general pattern of drainage in hill area is dendritic and sometime centrifugal and landslide and slumping are frequent. Churia goes maximum load to the river in the eastern and central Nepal.

6. **Terai:** This physiographic belt of Sapt Kosi river basin is the southernmost part of the Himalaya. It is the northern part of the Indo-Gangatic and extends from the east and west of Nepal. Widths varies from 10-30 km. the northern limit of this physiographic division is Churia or Shivalik hill, the general elevation of the Terai is very steep and rectilinear.

i) Elevation in this physiographic unit varies. It is 80 meter south of Chapa, 68 m. south of Krishna, 62 m. of east Jogbani and Hanumannagar, 88 m. It rises up to 152 m. The Terai region has lower elevation in comparison of Churia hill area. It can be say from the above analysis that Janakpur is lowest elevation in Terai.

ii) In the eastern Terai the gradient is 5 m./km and middle part 6 m./km and lower part 1.5m/km.

Terai region may be divided into three parts:

Bhabar zone: The bhabhar zone is 13 to 16 km wide and it is full boulder-gravel deposit. The Bhabhar is well developed in the major and medium rivers originating either Mahabharat or north whereas, it is poorly developed in the zones of rivers of the Churia hill.

Marshy Midland: Marshy Midland lies south of the Bhabhar zone. Most of the places spring line is found which creates the water logging in the middle Terai, which makes some marshy land.

Southern Terai: The land of Terai is practically plain in the middle and southern part, whereas due to Bhabar deposit it develops steep gradient in the north. Compared to eastern Terai, the western Terai is more undulation. North of Nepalganj the undulation is sharp. It is seen that in areas where Mahabharat river is flowing the undulation is absent. Erosion along with biological (human) forces had played its part to smoothen the topographic features.

The Plain: The plain part of Sapt Kosi river basin is an important relief unit, structurally, reason is a segment of great Indo-Gangatic track. However, it has some marginal partition of other major two formations that is the Shivalik in the northern part and the foothill area of the Himalaya. The foothill regions of the Shivalik formation covers a area of 364 sq. km an continues across the Indo-Nepal boundary towards north between the relatively lower peak 237 m near Santpur and the shorter and 32 km long 6-8 km wide. Ram Nagar doon hill range and the northern higher (Sumeshwar fort, elevation 865 m) and longer 72 km long. Sumeswar range lays Hars valley called doon about 22 km long and below 150 m in elevation that are three negotiable part. The Sureswar pass along the Jurepani stream, the Bhikhana thari pass up the Kudi river and the Marluood pass through the Harwa river valley. These streams have made the sedimentary formation very rough and rugged.

"It is evident that although the average feeding may not exceed 1,300 to 1,400 m, it varies width one part to another. Particularly, the depth increases up to 8 to 10 thousand m and the Himalaya are approached. There are two large troughs, which may call Darbhanga trough and Madhepura having 6,000 m. depth run almost parallel (Singh, 1965)." The plain part of Kosi river basin may be two broad regions (Singh, 1968).

Thus, the whole plain region is drained by Sapt Kosi and its tributaries, these rivers are responsible for feeling of with alluvial deposits of 2,000 m deep trough at the foot of Nepal Himalaya. They flow sluggishly in this flat land as a result of which the area is marked by local landmarks such as levees, bluff, oxbow lake, mashes, tal, reverine deposits etc. This Sapt Kosi river is very nutritious in the respect of provide to frequent floods.

DRAINAGE SYSTEM: 'Drainage refers to the manner in which precipitation falling in are or brought from outside is drained off (Ahmad, 1965).' Water, flowing down to the sea and immediately breath the bad surface breathing and ground water preside a bed for flowing streams and mass wasting may dump great quantities, river most carry all but small fractions of the total row waste from the land to the sea (Bloom, 2003).' Drainage is a natural plan on the surface through which all the forms of water received by an area flows downwards collecting all the water from the basin. The drainage characteristics of Sapt Kosi river basin has remained a very dynamic as well as an emerging issues worldwide.

'A drainage basin, alternatively described is a catchment area is in geomorphology and hydrology a region drained by a particular stream or by a river system (Fairbridge, 1968).' 'The size of the basin, total precipitation, and the loses due to evaporation and to absorptions by soil and vegetation (Fairbridge, 1968)' Drainage basin reflect the geomorphologist study of terrain, catchment area which is used more by hydrologist water supply by mind.

Drainage in any territory forms an integral part of the surface configuration forming two broad categories:

1. Surface water drainage and 2. The underground reservoir of the water.

Sapt Kosi river basin enjoys humid-tropical climate and high altitude mountain climate. Weathering is a dominant process of land sculpturing, drainage constitutes the most important element of surface geo-dynamics. Sapt Kosi river is an important tributary of Ganga forms by the confluence of seven rivers i.e. Sunkosi, Arun, Tamur, Indrawati, Bhote Kosi, Tama Kosi, and Dudh Kosi which originates in the snow cover peaks of Sikkim, Nepal and Tibet. They are fed by the glaciers of Gaurishankar-Everest-Makalu-Kanchanjunga group of peaks. First three tributaries contribute respectively 44%, 33% and 19% of the total water flow in the river uniting.

SURFACE WATER DRAINAGE: The whole drainage basin presents high undulation with tough terrain, on that basis the basin may be divided into two broad physiographic divisions.

a) Areas of high hills and mountains,
b) The plain track of Bihar.

It constitutes 68,944 sq. km (hill and mountains) and 22,532 sq. km (plains) most of the river gets origin from the upper part that is from the Tibet, the Great Himalaya, Middle Himalayan range as well as Shivalik range. There are three main types of rivers in the Sapt Kosi river basin that is i.e., Antecedent, Subsequent and consequent belonging to the different geological period governed by different orogenesis. Major rivers

TABLE - 2.2
DRAINAGE BASINS OF SAPT KOSI RIVER BASIN

Main River Basin	Total Area of River Basin	% of Main Basins of Total Area	Sub River Basin	Sub Basin's Area	% of Sub Basins of Total Area
	(In Sq. Km.)	(In %)		(In Sq. Km.)	(In %)
1. **ARUN RIVER**	34,660	38	a. ARUN RIVER	11,184	12.2
			b. YARU TSANGPO	11,003	12.2
			c. BUM CHU	6,281	6.86
			d. NIEN CHU	6,192	6.74
			TOTAL	**34,660**	**38**
2. **SUN KOSI RIVER**	16,743	18	a. DUDH KOSI	4,005	4.4
			b. SUN KOSI	3,857	4
			c. BHOTE KOSI	3,145	3.5
			d. TAMA KOSI	2,905	3.1
			e. INDRAWATY	2,831	3
			TOTAL	**16,743**	**18**
3. **TAMUR RIVER**	6,741	7		6,741	7
4. **SAPT KOSI**	33,332	37	a. SAPT KOSI	14,074	16
			b. BAGMATI	13,589	15
			c. KAMLA-BALAN	5,669	6
			TOTAL	**33,332**	**37**
TOTAL	**91,476**	**100**		**91,476**	**100**

Source: Computed From SOI Toposheet no. 72/G & 72K and Google Map.

which are now call system Sapt Kosi in the east. Present mountainous area was occupied by sea which was named as Tethys Sea. The antecedent river where existence before the Himalaya and the called their tributaries after the Himalaya and alpine orogenic, along the development of monsoon climate. The consequent rivers are developed after the shop and configuration of the area determine. The surface plan of water flow has been shown by the figure no. 2.2 and the table no. 2.2.

The tributaries of Sapt Kosi river basin in the mountainous track have got their origin after the evolution of different ranges and simply after the determination of reliefs (the Hill system).

The Arun River Basin: The Arun river basin is an important tributary of Sapt Kosi river, mostly flows in the China and Nepal territories. It lies between the latitude and longitude of 26^{0}99'49" to 29^{0}17'41" and 85^{0}67'06" to 89^{0}18'72" respectively. The part of the basin falls under the China is 30,868.9 sq. km and under the Nepal 3,791.1 sq. km. Total length of Arun river is 370 km. Arun river is an antecedent drainage basin because it has retained its flow regime even after the evolution of the different ranges and peaks of Himalaya. It flows eastward and then turns southward and joins the main stream of Sapt Kosi river at Tribeni. Total basin area of Arun river is about 34,660 sq. km. Arun river basin is separated from Brahmputra (Yarlung Tsangpo) river basin in the northern side. It is evident from the above that Arun river has 40 tributaries in which three river viz. Yaru Tsangpo, Bum Chu and Men Chu in the south Tibet (China) forms 434 km length and 25.8 % of the total area of the Arun river basin. The remaining 37 small drainage basins and tributaries of the Arun river out of which 13 comes under Chinese territory whereas 24 drainage basin comes under Nepal Himalayan territory and covers 11,184 sq. km i.e. 12.2 % of the total area of the Arun river basin.

The Arun is the largest trans-Himalayan river passing through Nepal and also has the greatest snow and ice covered area of any Nepalese river basin. The Arun drains more than half of the area contributing to Sapt Kosi river system but provides only about a quarter of the total discharge. This apparent contradiction is caused by the location of more than 80 % of the Arun's drainage area in the rain shadow of the

Himalayas. Average annual rainfall in Tibet is about 300 mm (Kattelmann, 1990). Arun river basin covers 1/5 area of Sapt Kosi river basin. It is 12.2 % of the total area of the basin (Table no. 2.3).

Arun river basin: Southern part of Tibet is mostly covered and drained by the Arun river which falls in northern part in Kosi river basin. The whole river regime of the Arun river in Nepal

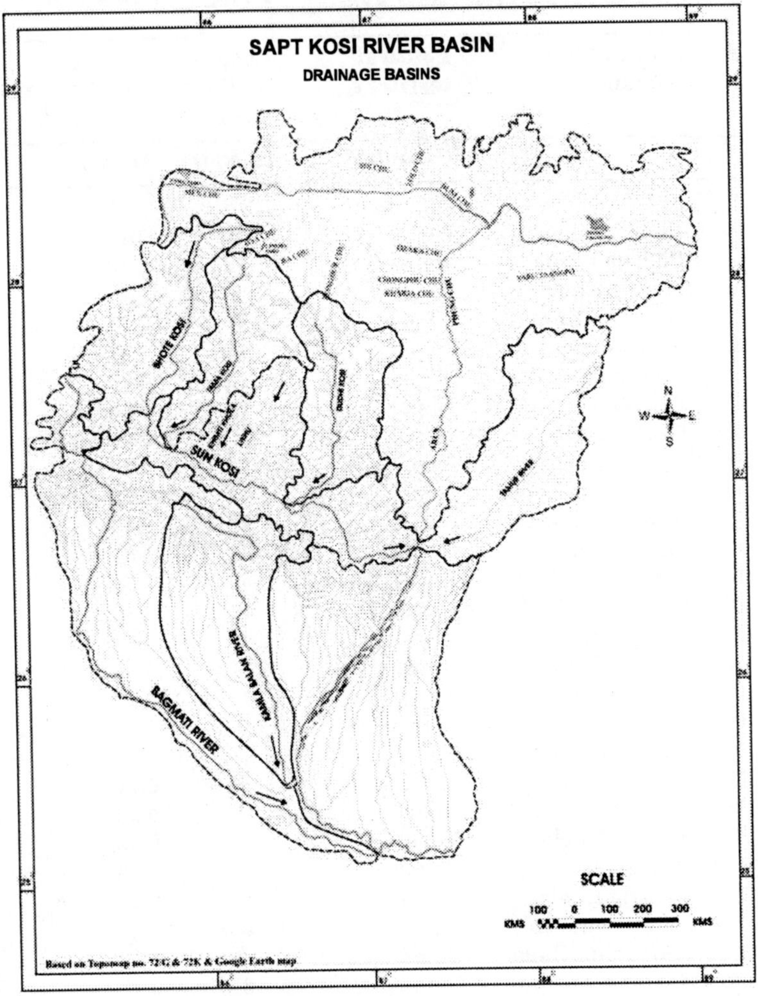

Fig. 2.2

and China territory flows through mountain tract. The altitude of the Arun river ranges from highest and in the higher Himalaya 2,248 m. near Kimathangka La and the minimum in the Shivalik range having altitude of 183 m. in Chatra. In whole regime i.e. in the lower, middle and upper reaches the course of the river form very sinuous as well as meandering.

Table - 2. 3
ARUN RIVER AND THEIR TRIBUTARIES

S. No.	Name of the Stream	Political territory (Country)	Length of the stream (in km)	% of the Total stream Length (in %)
1.	Arun	China-Nepal	370	8.01
2.	Dhanre Khola	Nepal	6	0.13
3.	Pikhwa Khola	Nepal	23	0.49
4.	Munga Khola	Nepal	5	0.11
5.	Mangma Khola	Nepal	8	0.17
6.	Leguwa Khola	Nepal	13	0.28
7.	Chhintalung Khola	Nepal	15	0.32
8.	Piluwa Khola	Nepal	25	0.54
9.	Hinwa Khola	Nepal	2	0.04
10.	Sabhyaya khola	Nepal	10	0.22
11.	Chirkhuwa khola	Nepal	13	0.28
12.	Irkhuwa Khola	Nepal	20	0.43
13.	Sinwa Khola	Nepal	42	0.91
14.	Sankhuwa Khola	Nepal	42	0.91
15.	Chhoyang Khola	Nepal	12	0.26
16.	Iswa Khola	Nepal	2	0.04
17.	Kusuwa Khola	Nepal	11	0.24
18.	Maghang Khola	Nepal	8	0.17
19.	Wabak Khola	Nepal	8	0.17
20.	Barun Khola	Nepal	28	0.61
21.	Dima Khola	Nepal	5	0.11
22.	Chhong Khola	Nepal	15	0.33
23.	Bagang Khola	Nepal	10	0.22
24.	Tiju Khola	Nepal	13	0.28
25.	Wakang Khola	Nepal	5	0.11
26.	Karmachu Khola	China	8	0.17
27.	Phungchu Khola	China	10	0.22
28.	Naktangchu Khola	China	8	0.17
29.	Men Chu	China	91	1.98

30.	Nien Chu	China	63	1.36
31.	Gya Chu	China	35	0.76
32.	Ra Chu	China	56	1.22
33.	Rong Buk Chu	China	50	1.08
34.	Dzaka Chu	China	35	0.76
35.	Shi Chu	China	42	0.91
36.	Gyanor Chu	China	35	0.76
37.	Phung Chu	China	56	1.21
38.	Lolo Chu	China	50	1.08
39.	Bum Chu	China	154	3.33
40.	Yaru Tsangpo	China	189	4.09
41.	Kharta Chu	China	32	0.69
		TOTAL	**1625**	**35.17**

Source: Computed from Google earth and collected from Rivers of Nepal, Sharma, K.C., 1997.

The Arun river basin: The Arun river which is the upper part of the Sapt Kosi river basin flows through high mountain and hills forming deep gorges of Yo Ri gorge (1,829 m. deep). The important gorge of the basin is Chatra gorge (10 km long and 5-8 km wide). Arun is a perennial river characterised by dendritic pattern but it is also characterised by Radial, Pinnet and Herringbone pattern. Dzaka chu, Chongphu chu, Kharta chu and Rong buk chu are major tributary of Arun river which originates from Mt. Everest's north and south side's glaciers. Kimathangka La (2,248 m.) is located near Nepal-China boarder along the Arun river valley.

Yaru Tsangpo river basin: This basin is located in extreme north eastern frontier of Sapt Kosi river basin. A big lake Tsomo Chamling (4,739 m) is situated in this basin, which is about to freezes during whole winter season and melts in summer. This basin is mostly covers Tibet plateau. Gampazhoi is a major Tibetian city. Yaru Tsangpo river basin is extended over east to westward direction in about 11,003 sq. km area. Total length of Yaru Tsangpo river is 189 km.

Nien chu basin: This basin is located in extreme north-western corner of Sapt Kosi river basin. Major tributaries are Gya chu, Men chu and Ra chu. Ra chu originates from northern side of Gauri Shankar peak's glaciers. Nien chu basin is extended over west to eastward direction in about 6,192 sq. km area. Total length of Nien chu river is 63 km. One major lake is located in extreme north-western corner of the basin known as Langjang Tsho. The river basin is lies over Tibet plateau possessing height of averagely more than 4,000 m. Most of drainage network is followed by dendratic pattern.

Bum chu basin: After Nien chu basin the river is known as Bum chu basin, which is extended up to Arun basin in the eastern south direction. Near Silin village (elevation more than 4,000 m.) Bum chu gets southward turn and meets with Phung Chu. Bum chu river basin covers area 6,281 sq. km of Sapt Kosi river basin, which is 6.86 % of the total area of the basin.

Sun Kosi river basin: Sun Kosi river is an important tributary of Sapt Kosi river basin flows throughout the Nepal. It is constituted by five main river channels i.e. Indrawaty, Bhote Kosi, Tama Kosi, Dudhi Kosi and Belphi Khola.

Dudh Kosi river basin: Dudh Kosi is an important tributary of Sun Kosi flows north-south at upper part and parallel to Likhu at lower part. The valley of Dudh Kosi is nearly 70 mile long. The course is highly meandering in the lower part. Few high altitude terraces are found in the valley.

Sun Kosi river basin: It covers 3,857 sq. km of the total area of the basin. The Sun Kosi with its five main tributaries with a numerous small finger tips channels of first order (table no. 2.4). The main channel of Sun Kosi river basin is perennial. It's important tributaries are also perennial. Most of the finger tip channels become dry in summer. Due to retreat of snow from the upper reaches of this river the hydrologic regime has been badly affected. This river flows due east parallel to the Mahabharat range of the Shivalik.

Table - 2.4

SUNKOSI RIVER AND THEIR TRIBUTARIES

S. No.	Name of the Stream	Political territory	Length of the stream	% of the Total stream Length
		(Country)	(in km)	(in %)
1.	Sunkosi	Nepal	270	5.85
2.	Sarong Khola	Nepal	12	0.26
3.	Yari Khola	Nepal	15	0.33

4.	Rankua khola	Nepal	5	0.11
5.	Laikhu Khola	Nepal	6	0.13
6.	Chhalang khola	Nepal	8	0.17
7.	Dunkhuwa Khola	Nepal	8	0.17
8.	Sawan Khola	Nepal	5	0.11
9.	Bhote Kosi	Nepal-China	140	3.03
10.	Chakhu Khola	Nepal-China	8	0.17
11.	Bhairavkhund Khola	Nepal	6	0.13
12.	Libling Khola	Nepal	12	0.26
13.	Jum Khola	Nepal	4	0.09
14.	Belephi Khola	Nepal	55	1.19
15.	Jhikhu Khola	Nepal	3	0.07
16.	Nosem khola	Nepal	5	0.11
17.	Duskol Khola	Nepal	7	0.15
18.	Risama Khola	Nepal	8	0.17
19.	Indrawati Khola	Nepal-China	59	1.28
20.	Chunari Khola	Nepal	42	0.91
21.	Khani Khola	Nepal	35	0.76
22.	Malemchi Khola	Nepal	37	0.81
23.	Sindhu Khola	Nepal	24	0.52
24.	Rosi Khola	Nepal	15	0.33
25.	Tamba Kosi	Nepal	80.5	1.74
26.	Bhatauli Khola	Nepal	10	0.22
27.	Milti Khola	Nepal	8	0.17
28.	Khimti Khola	Nepal	33	0.71
29.	Bhotekosi	Nepal	5	0.11
30.	Likhu Khola	Nepal	80.5	1.74
31.	Dudhi Kosi	Nepal	170	3.68
32.	Thotne Khola	Nepal	3	0.06
33.	Rawa Khola	Nepal	5	0.11
34.	Solu Khola	Nepal	6	0.13
35.	Inkhu Khola	Nepal	28	0.61
36.	Hingu Khola	Nepal	10	0.22
37.	Lumading Khola	Nepal	12	0.26
38.	Kyanghar Khola	Nepal	4	0.09
39.	Imja Khola	Nepal	21	0.45
40.	Legmoche Khola	Nepal	13	0.28
41.	Thami Khola	Nepal	16	0.35
42.	Bhote kosi	Nepal	4	0.09
43.	Dudh kosi	Nepal	3	0.05
44.	Cholakhu Khola	Nepal	5	0.11
		TOTAL	**1306**	**28.29**

Source: Computed from Google earth and collected from Rivers of Nepal, Sharma, K.C., 1997.

Bhote Kosi river basin: The Sun Kosi river meets Likhu and Ulanzeb making bed travels south-western and receives Dudh Kosi at Kurule. Dudh Kosi gets its origin from Mt. Everest region and flows due south west. From Kurule and track south angle bend at kempughtal, onwards it flows to meet the Arun and Tamur river at Tribeni ghat. It has the length of 170 km. the Belphi flows more or less north-south parallel to Sun Kosi. It has wide valley with a suspension bridge at Jalbire. On the western slope peak Chaktara Bazar located. It drains Chang, Samarphu (3,952 m.) and Bharmkund lake which is located at watershed of Sun Kosi and Balephi. Indrawaty flows north to south mostly sinuous and very narrow at Sipaghat moving easterly bend. Its main tributaries are Malemchi Khola, Jayarani Khola and Chowk khola. Nyalam (3,750 m) is a major Chinese city just near Nepal border whereas Kodari (2,300 m) and Brahbise (1,220 m) is located in Nepal beside Bhote Kosi river.

Indrawaty river basin: The Indrawti River Basin is located in Bagmati zone, Central Development Region of Nepal. It originates from the Mahabharat and Himalayan Ranges at an elevation of about 5,850 m above mean sea level (msl) and discharges into the Sun Kosi River at an elevation of 626 m msl. Both snowmelts and spring sources contribute to the base flow of the river. The total length of the Indrawati River from its origin till the confluence with Sunkosi is about 59 km and the catchment area is about 1,240 km^2. The basin extends from 27° 37' 11" N to 28° 10' 12" N latitudes and 85° 45' 21" E to 85° 26' 36" E longitudes. The Indrawati Basin covers approximately 43% area of Sindhupalchowk district and 11% area of Kavrepalanchowk district. The lower reach of the river basin such as the Melamchi river (at Talamarang), Nayagaun (via Nagarkot and Kunta Besi) and Dolalghat has vehicular access. Indrawati River receives flow contributions from the following major tributaries: Larke, Yangri, Melamchi, Jhyangri, Jhyangri, Chaa, Handi, Mahadev (Sharma, 1997).

Table - 2.5

TAMUR RIVER AND THEIR TRIBUTARIES

S. No.	Name of the Stream	Political territory	Length of the stream	% of the Total stream Length
		(Country)	(in km)	(in %)
1.	Tamur	Nepal	150	3.25
2.	Kabeli Khola	Nepal	44	0.95
3.	Mewa Khola	Nepal	5	0.11
4.	Kosini Khola	Nepal	6	0.13
5.	Sinse Khola	Nepal	10	0.22
6.	Telis Khola	Nepal	14	0.31
7.	Nabhu Khola	Nepal	9	0.19
8.	Sanwan Khola	Nepal	8	0.17
9.	Khorang Khola	Nepal	12	0.26
10.	Koya Khola	Nepal	10	0.22
11.	Nibhu Khola	Nepal	28	0.61
12.	Hima or Phemi Khola	Nepal	12	0.26
13.	Maewa khola	Nepal	6	0.13
14.	Mewa Khola	Nepal	4	0.09
15.	Tamur Khola	Nepal	10	0.22
16.	Phesua Khola	Nepal	13	0.28
17.	Dhobi Khola	Nepal	3	0.07
18.	Phongsuwa Khola	Nepal	12	0.26
19.	Leoti khola	Nepal	15	0.33
20.	Dhankuta Khola	Nepal	5	0.11
21.	Tankhuwa Khola	Nepal	5	0.11
22.	Raghuwa khola	Nepal	10	0.22
23.	Chharuwa Khola	Nepal	12	0.26
		TOTAL:	**403**	**8.76**

Source: Computed from Google earth and collected from Rivers of Nepal, Sharma, K.C., 1997.

Tama Kosi river basin: Tama kosi has 50 mile length and flows due south-west. It has wide course with easterly bend. The main tributaries are Chanaty khola, Khimti khola, and Bhote Kosi. The valleys wide near the upper part and narrow down in the south. Likhu khola has also about 50 m channel. Its valley comparatively narrow and steep.

Tamur River basin: Tamur river basin lies mostly in Nepal territory covering 6,741 sq. km area that is i.e. 7 % of the total area of Sapt Kosi river basin. It lies between the latitude and longitude of $26^{0}97'24"$ to $28^{0}26'88"$ and $87^{0}13'35"$ to $88^{0}54'73"$ respectively. Tamur main stream originates from Kanchanjunga glacier and flows due south-west to meet Jogwa khola at Saka. The eastern bend is at Taplejung and receives Kabeli khola which comes from the east. The river is highly sinuous and meandering characterised by numerous meanders. It has also been dammed at several places and also characterised by naturally interlocked lakes. Tamur river meanders a while at Angbung and from there it become south-westerly to westerly and join by other tributaries of Sapt Kosi river at Tribeni. The total length from its source to north of the river is 150 km. from Tribeni onward the same river is called Sapt Kosi and flows straight south. It's come out at Churia hill at Chatra and from then it makes south-westerly bend. During summer the adventeceous streams which are not fed by the permanent glacier becomes dry, but main stream of the Tamur originated from Kanchanjunga, which is extremely frozen. Supplies water throughout the year and hence is a perennial river. Owing to high gradient of slope the erosionablity of this river is very high and its powers large quantity of sediments in Sapt Kosi river from its regime. Before entering into the Indian territory. Sapt Kosi travels nearly 16 km in Nepalese territory from the point of Tribeni. The paleao-channels reveal the fact that this river flows 200 years back near Purnea, now it flows 160 km west from the original course. During Monsoon season when the whole river regime receive high amount of the percentage of annual rainfall, carries water through the sinuous course of the upper reaches whole it became extensive while entering the plain track of the Indian territory. Owing to the high gradient and the speedy discharge of water it carries high amount of sediments, which might be the course of the shift of the main channel (Table no. 2.5).

Sapt Kosi river basin (upper): After Tribeni (meeting place of Sunkosi, Arun and Tamur) Kosi river is know n as Sapt Kosi because of meeting seven tributary (Indrawaty, Bhote Kosi,

Tama Kosi, Dudh Kosi, Sun Kosi, and Arun, Tamur). Sapt Kosi river basin includes three big river basins i.e., Bagmati, Kamla-Balan, and himself Sapt Kosi.

Sapt Kosi river basin (lower): Sapt Kosi has been named after seven rivers of the Kosi river system of the plain tract of Indian subcontinent. Sapt Kosi in Nepal form by the Sun Kosi of the western, Arun of the central Nepal and the Tamur in the eastern Nepal, when Kosi enters in the plain region of the Gangatic belt forms only a single flow as a Sapt Kosi main stream except the Kamla-Balan and Bagmati which are the thyea tool tributary in the plain tract of Sapt Kosi river basin (Table no. 2.6).

In the plain these river covers an extensive area 14,074 sq. km constituting 16 % of the total area of the basin. Owing to the less gradient the river water of Kamla-Balan and Bagmati along with the Sapt Kosi river spreads over an extensive area forming a huge plain tract. In the plain tract the Sapt Kosi river constitute of only three important tributaries i.e. Bagmati, Kamla-Balan, Tilyuga and the main stream of Sapt Kosi.

Bagmati river basin: Bagmati river basin get its origin from the southern range of Sheopurilekh, north of Kathmandu basin and flows straight south-west cutting the Mahabharat range. The drainage area is 13,589 sq. km that is 15 % of the basin. It appears that the present nature of river is comes into existence Pleistocene time when Kathmandu lake disappeared. Bagmati flows to south-west from Kathmandu up to Jhanalkot from there to Gangate in the south and later on enters the Indian territory making 'U' bend at Betehani. Upto Hariharpurgarhi to south-east direction and finally to the south. This river is also perennial in nature but most of its finger tips stream lies away during the summer. This river gets flooded and in the lower tract, while the upper tract is mountainous. The steep gradient resultant into high rate of discharge, erodiblity as well as carrying capacity of sediments. It also forms a fan shaped deposited landform below the foothills, an extensive and huge deposition is found in the Terai region of Bihar. It carries high amount of sediment from the upper reaches.

The important tributaries of Bagmati is Martin khola having length of 52 km, Jhal Nadi (74 km) and Chandi khola has the total length of 38 km and along with these main tributaries a number of small finger tip streams are also found in the basin.

Kamla-Balan river basin: Kamla and Balan joining together have been named as Kamla-Balan, an important tributary of Sapt Kosi river basin in the plain region of the Gangatic belt. It originates from Sindhuligarhi and flows to south making easterly trend parallel to Churia hill and flows due south and joins the main stream of Sapt Kosi in the lower Gangatic plain located in the end of the basin. The total area of the basin covers 5,669 sq. km constituting 6 % of the total area. This river is highly migratory due to the heavy sediment load brought by upper reaches of the river an extensive alluvium track has been formed by three rivers. This river seems like a herringbone pattern of stream in the piedmont area where it is of dendritic pattern as a whole. Thus, Sapt Kosi river which receives waters of all tributaries existing in the upper part catchment area of the basin.

'The river Kamla rises in Nepal, enters Darbhanga district just east of Jaynagar town in the north and flowing southwards and then in a south-easterly direction to Tilkeshwar falls into Tiljuga and ultimately into Kosi by various tortuous routes. This river changes its course from year to year and its old beds are found all over the north of Madhubani subdivision. The main channel used to flow ten miles east of Madhubani, then ten miles west and now it flows 2 miles east of Madhubani town and reaching Ryam encircling it by various channels takes ultimately the bed of Gause river and then goes south eastward to Tilkeshwar. It is a large river in the rains and is liable to heavy floods (Karki, 2005).'

Table - 2.6

SAPT KOSI RIVER AND THEIR TRIBUTARIES

S. No.	Name of the Stream	Political territory (Country)	Length of the stream (in km)	% of the Total stream Length (in %)
1.	Bagmati	Nepal-India	597	12.93
2.	Martin Khola	Nepal	52	1.13
3.	Jhal nadi	Nepal	74	1.6

4.	Chandi Khola	Nepal	38	0.82
5.	Kamla-Balan	Nepal-India	328	7.11
6.	Tilyuga	India-Nepal	11	0.24
7.	Baiti	India	8	0.17
8.	Purain	India	5	0.11
9.	Soharain	India	6	0.13
10.	Dhemura	India	9	0.19
11.	Tilabe	India	8	0.17
12.	Soneh	India	8	0.17
13.	Parwane	India	10	0.22
14.	Karuwa	India	12	0.26
15.	Waiwah	India	12	0.26
16.	Chilauni	India	5	0.11
17.	Hiran	India	12	0.26
18.	Haha	India	9	0.19
19.	Lakcha	India	8	0.17
20.	Libri	India	10	0.22
21.	Kalikosi	India	10	0.22
22.	Pannar	India	12	0.26
23.	Suguruce	India-Nepal	12	0.26
24.	Tilabeh	India	13	0.28
25.	Dhusan	India	2	0.04
26	Sursar	India-Nepal	12	0.26
		TOTAL	**1283**	**27.78**

Source: Computed from Google earth and collected from Rivers of Nepal, Sharma, K.C., 1997.

CLIMATE: Climate influences the content the geomorphological processes, hence the assessment the climatic characteristics of the region becomes an integral part of the geomorphic study. The advocate of the climatic geomorphology argue that the rate of epation of weathering and erosion processes, vegetation type surface runoff nature and rate of erosion and mechanism of landform genesis and development area the mean theme of the geomorphic study (Roy, 1964).

Hence, it is notable that the area divide of vegetation are directly pelted by falling raindrop with maximum kinetic energy and causes splash erosion where in loose partials are resulted on the ground surface and form strong cuirasses which discourse infiltration of rain water and form increased surface runoff on the other hand densely vegetated area mainly of forest are characterised by least splash erosion because of maximum interception of rain water (Stoddart, 1969).

The climate of a region directly inveals in geomorphic processes leading to the development of vast featureless plain and the landform development.

The climate of Sapt Kosi river basin ranges from the tropical south to freeze north. The tropical zone covers the whole Gangatic plain, south of Janakpur-Baraha-Chatra, the subtropical zone covers the surrounding area along the Sino-Nepal highway and Zhangmu,

Temperate zone covers the area along the line from Zhangmu to Nyalam and the freeze zone covers the areas above the Nyalam mountain pass. The element of the weather the factors controlling the climatic condition of Sapt Kosi river basin is the Himalaya. Precipitation is very uneven distributed throughout the basin; the annual precipitation is 300-400 mm in the northern Himalayan region, 1,000 to 1,500 mm in subtropical and tropical region and 1,500 to 2,500 mm in the temperate region. There are high variations in the weather conditions throughout the basin (Table no. 2.7).

Table - 2.7

REGIONAL DISTRIBUTION OF CLIMATIC CONDITION OF SAPT KOSI RIVER BASIN

S.N.	Climatic Zone	Area under each Climatic Zone (in Sq. Km.)	Percentage (%) of each Climatic Zone
1.	Sub-Tropical Monsoon (Am)	56,010	61
2.	Temperate	32,631	36
3.	Freeze	2,835	3
TOTAL		91476	100

Source: Calculated from topomap no. 72/G & 72K of Indian side and Google Earth for Nepal and Chinese side.

Thus, the whole river basin interfaces with diverse climatic conditions. As some part of the basin extremely high annual mean evapotransporation where as other part of suffers frequent drought and soil erosion. The temperate zone has vast

Table - 2.8

SAPT KOSI RIVER BASIN: ANNUAL RAINFALL

S. No.	RAINFALL (in mm)	COUNTRY	JAN	FEB	MAR	APR	MAY	JUN	JUL	AUG	SEP	OCT	NOV	DEC	YEARLY MEAN
			MONTHS												
1.	GODAVARI (2011)	NEPAL#	4.2	47.6	27.7	62.5	345.2	402.6	417.8	399.1	298.4	0	0	0	167.1
2.	JANAKPUR AIRPORT (2012)	NEPAL#	20.3	4.4	6.2	45.1	44.4	162.4	140.4	373.6	199.9	45.5	0	0	86.8
3.	JIRI (2012)	NEPAL#	16.2	25.1	12	82.3	73	245.4	654.9	722.8	459.1	25.4	8.2	0	193.7
4.	KATHMANDU AIRPORT (2012)	NEPAL#	11.5	45.4	27.3	44.5	278.6	299.1	428.5	451.4	217.3	95.7	0	0	158.3
5.	OKHALADHUNGA (2012)	NEPAL#	6.2	10.2	5.5	86.8	164.2	294.6	367	484.6	299.8	2.4	0	0	143.4
6.	PHATEPUR (2011)	NEPAL#	3.3	18	31	117.2	143.7	468.3	328.9	520.8	144.3	4.5	70.9	0	154.2
7.	CHATRA (2013)	NEPAL#	4.5	31.7	20.1	65.7	170.7	480.7	467.4	388.4	207.9	193.7	25.2	0	171.3
8.	DHANKUTA (2013)	NEPAL#	8.6	37.8	7	91.9	155.6	184.1	264.9	51.7	48.8	131.6	0	0	81.8
9.	SAHARASA (2013)	INDIA*	0	10.6	3.2	132.4	22.4	191.6	225.5	173	285.8	144.6	0	0	99.1
10.	SAMASTIPUR (2013)	INDIA*	2	7.6	0	8.3	35.8	109.7	127.8	95.8	144.6	214.5	0	0	62.2
11.	SUPAUL (2013)	INDIA*	11.8	9.4	3.8	17	65.1	335.6	308.4	177.9	191.1	171.5	0	0	107.6
12.	MUZAFFARPUR (2013)	INDIA*	16.1	43.9	0	25	149.5	161.1	79.9	122.5	89.4	230.7	0	0	76.5
13.	MADHUBANI(2011)	INDIA*	11.9	3.4	43.8	243.5	132.3	251.7	159.8	249.2	7.9	1.7	0	0	92.1
14.	MADHEPURA(2013)	INDIA*	3.7	7	0.3	17.6	93.6	374.4	199.9	197.3	224.9	302	6	0	118.9
15.	DARBHANGA (2011)	INDIA*	10.1	26.3	19.7	136.7	155.9	312.5	210.6	376.7	13.4	0.5	0	0	105.2
16.	SITAMARHI (2013)	INDIA*	27	27.2	0	30.9	109.2	71.6	93.2	140.3	115.4	155.2	0	0	64.2
17.	NYALAM (Average)	CHINA**	12	2	12	21	51	93	132	132	90	18	0	12	47.9

Source: a) #Nepal - DHM, Kathmandu, b) *India - IMD, Pune and c) **China - http:/www.worldweatheronline.com (for Nayolam), downloaded on 24-4-2014 vegetation cover and clear correlation between elevation and precipitation with the maximum precipitation at Jhangmu (2,200 m MSL).

It is evident from table no. 2.8 that the tropical monsoon (Am) which is found in the plain region of Indian subcontinent covers an area of 56,010 sq. km i.e. 61 % of total basin. Temperate climatic zone covers an area of 32,631 sq. km i.e. 36% of the total area of the basin. It is also called as Highland Mountain (H) and characterised by conical trees with increasing human interference and settlement, whereas, the high altitude freeze climate is found in the higher altitude covering an area of 2,835 sq km i.e. 3% of the total area. The highest rainfall in the basin reached at Iri (193.7 mm yearly mean), where as the minimum rainfall found to occur at Nyalam (47.9 mm yearly mean). The average annual rainfall in the whole Sapt Kosi river basin is 113.55 mm, where as it varies significantly in the plain tract of India, the valleys and mountains of Nepal and Chinese territory. The monsoon side face of the Himalaya receives comparatively higher amount of rainfall where in leeward side of the Chinese territory receives relatively very less amount of rainfall and this part of the territory receives low and of rainfall. Most of the finger tip streams gets there origin from the middle and higher Himalaya. The streams of the middle and Nepal Himalaya receives moderately high rainfall were as the plain tract receives comparatively low amount of rainfall than the middle Himalaya. A record of rainfall of 50 years average shows that there are high fluctuations the amount of rainfall falling in area. This variations leads to the variations in the development of landform in the basin. The face ward side like the Shivalik, Nepal Himalaya receives comparatively high amount of orographic rainfall.

From the above table no. 1.11 it is apparent that all the stations of the basins having variations in the amount of rainfall they receive. The amount of rainfall which flows through the streams of the basin has seriously eroded the river regime. The access flows of water during the south-west monsoon in the stream have course heavy severe erosion in the high slope area and heavy siltation in the lower catchment area of the basin.

TEMPERATURE: Sapt Kosi river basin lies in three distinct temperature zone that the sub-tropical, temperate and polar type freezed zone where temperature correlation quit varies. In the plain tract of lower catchment area of Sapt Kosi river basin experiences seasonal reversal of weather element will high temperature during the summer and low temperature during the winter whereas the sub-tropical and low temperate regions have low temperature throughout the year and this region having no variation in the amount of annual range of temperature. Temperature remain below 0^0C in high altitude freezed zone and sometimes below -50^0C in the higher altitude freezed zone but this region fluctuates as the lower most part the freezed zone has comparatively high temperature than the more altitudinal area of the same zone. Fluctuation in the temperature leads to melting of snow. The stream of Sapt Kosi river basin receives water from the melting of the snow and glaciers.

Temperature (Table no. 2.9) is the significant climatic parameter, which not only influences but also controls the different geomorphometric processes. In Sapt Kosi river basin temperature varies considerably in different climatic regions. There is frequent and wide spread frosting in the region of freezed zone. It is mostly found in the high altitudinal areas where derived freeze and they leads to alternate process of contraction and expansion. There is also conjelifraction in the high altitudinal areas. In the mountainous and intermountain areas temperature also has plays a significance role. There is large daily range of temperature in Table no. 2.9 which leads to exploitation, flacing and onion weathering in the plain region of Bihar their seems seasonal reversal in temperature condition, high temperature (high mean maximum temperature) is higher in the month of May-June where as it lower during the winter. This rhythm is not found in the temperate and high mountainous region.

Other climatic parameter like humidity, precipitation, windblown in an area also affects the basin. During winter all the intermountain area valleys remains invisible duw to fog and hell. High humidity is found during the rainy season. It is also slidly high during the month of December and January. Orographic rain also affects the whole basin. The basin is also experiences high wind speed during some particular month in

Table - 2.9

SAPT KOSI RIVER BASIN: ANNUAL TEMPERATURE

S.N.	STATION	COUNTRY	TEMPERATURE (in °C)	J	F	M	A	M	J	J	A	S	O	N	D	YEARLY MEAN (in °C)
01.	GODAVARI (2013)	NEPAL	Mean minimum	0.2	4	7.5	9.7	13.4	16.2	15.4	15.1	13.4	11.2	4.3	1	9.28
			Mean maximum	17.6	20	24.2	25.8	26.7	26.4	26.2	25.8	26.4	23.5	20.6	17.9	23.4
02.	JANAKPUR AIRPORT (2012)	NEPAL	Mean minimum	9.5	10.6	14	21	24.5	26.7	26.9	27	26.1	21.1	13.3	11.2	19.33
			Mean maximum	21.2	26.4	31.1	35.4	37.6	35.2	33.3	33.9	33	32.2	28.9	21.4	30.8
03.	JIRI (2012)	NEPAL	Mean minimum	-3.3	-0.5	2.5	7.1	11	15.9	17.9	16.8	15.2	8.4	2.3	-0.5	7.74
			Mean maximum	13.5	17.8	20.1	22.7	24.9	25.6	24.2	24.8	24.5	21.7	18.2	16.6	21.22
04.	KATHMANDU AIRPORT (2013)	NEPAL	Mean minimum	1.6	6.1	10.1	12.4	17.1	20.1	20.4	20.2	18.9	15.9	7.7	4.8	12.94
			Mean maximum	18.9	23.8	28.1	29.3	30	30	29.4	29.6	29.8	26.6	22.6	18.8	26.41
05.	OKHALADHUNGA (2012)	NEPAL	Mean minimum	3.1	6.3	9.2	11.2	13.2	15.2	15.2	15	14.2	10.4	6.2	3.4	10.22
			Mean maximum	15.2	19.1	21.9	24.3	26.9	26.2	25.3	25.9	23.6	24	21.8	18.6	22.74
06.	PHATEPUR (2011)	NEPAL	Mean minimum	6.7	10.9	14.7	17.4	21.7	22.8	23.7	23.9	23.7	20.3	14.7	9.8	17.5
			Mean maximum	23.2	26.6	32.1	34.4	33.8	34.4	34.4	34.3	34.4	34.3	31	26.9	31.6
07.	DAMAN (2009)	NEPAL	Mean minimum	5.4	5	7	10.7	12.4	14.9	17.2	16.4	16.7	13.1	7	3.1	10.74
			Mean maximum	14.4	17.5	17.8	22.1	22.2	23.8	24.3	23.5	24.8	22.5	18.9	14.5	20.5
08.	SAHARASA*	INDIA	Mean minimum	9	12	17	22	24	27	25	26	26	22	15	10	19.58
			Mean maximum	22	26	32	37	36	36	32	33	32	32	29	24	30.92
09.	SAMASTIPUR*	INDIA	Mean minimum	9	12	16	22	25	26	25	26	25	21	15	10	19.34
			Mean maximum	22	26	32	37	38	35	32	33	32	31	28	24	30.84

10.	SUPAUL*	INDIA	Mean minimum	9	12	17	22	25	27	26	26	26	21	15	10	19.67
			Mean maximum	22	26	32	37	38	36	33	33	32	31	29	24	31.1
11.	MUZAFFARPUR*	INDIA	Mean minimum	9	12	17	22	25	27	26	26	26	22	15	10	19.75
			Mean maximum	22	26	32	37	38	36	33	33	32	32	28	24	31.1
12.	MADHUBANI*	INDIA	Mean minimum	9	12	17	22	25	27	26	26	26	22	15	10	19.75
			Mean maximum	22	26	32	37	38	36	33	33	32	32	29	24	31.17
13.	DARBHANGA*	INDIA	Mean minimum	9.2	11	15.1	19.4	21.2	22.9	23.8	24.2	23.8	21.2	15.8	10.6	18.2
			Mean maximum	23.1	25.8	31	35.1	35	34.9	32.5	32.8	32.5	31.6	29	24.8	30.7
14.	SITAMARHI*	INDIA	Mean minimum	9	12	17	22	25	27	26	26	26	21	15	10	19.67
			Mean maximum	22	26	32	37	38	36	33	33	32	31	29	24	31.1
15.	NYALAM (Average)	CHINA	Mean minimum	2	4	7	10	13	16	17	16	15	12	7	3	10.17
			Mean maximum	16	18	21	24	25	25	24	25	24	23	20	17	21.84

Mean minimum temperature of Kosi river basin	15.59	Mean Temperature of Kosi river basin	21.65^{0}C
Mean maximum temperature of Kosi river basin	27.69		

Source : a) For Nepal portion - DHM, Nepal; b) For Indian portion - IMD, India and c) For Chinese portion - http://www.worldweatheronline.com/ (for Nyalam), *http://www.myweather2.com/(for Indian Portion) Downloaded on 24/04/2014.

excess and comparatively low in the month of the year. Thus, the local wind like Loo and Westerly also affect the river basin during the summer season. Loo a local wind heavily effect the region as it carries large amount of sand partials to form various features in the Sapt Kosi river basin.

NATURAL VEGETATION: Natural vegetation is the canopy of plants on the surface of the earth. The climate conditions are such that left to itself the gree cumered by forest, trees in course of time. Owing to an extensive area abrupt difference in the altitude has resulted into very conistrust natural vegetation throughout the basin. This is the basin where the lofty world highest altitudinal hill of the Himalaya (Mt. Everest 8,848 m.) lies. In the Sapt Kosi river basin minimum altitude above the MSL is 47 m. at confluence of Kosi-Ganga river near in Kursela. In the plain tract of Bihar tropical dry deciduous vegetation occurs where as in most part of the Shivalik, Middle and the Higher Himalayas characterised by the low temperature climate conditions yielding temperate conical, evergreen trees. The high altitudinal freezed zones are devoid of the natural vegetation only to some extend Mosses and Likens are found to occur.

Tropical dry deciduous vegetation has the following characteristics:

i) It occurs in area less than 15 inch of annual rainfall.

ii) It includes dry Sal forest but apart from Sal are other important species.

iii) Forest of the Sumeswar and Doon ranges in the Madhubani, Darbhanga, Supaul district constitute moist deciduous type of forest vegetation. Scrubs and nubs of grasses and redds. Yet, the sub Himalaya foothill forest are deciduous in nature, the permanent trees are Sal (Soria rawasta), Shisham (Dalbruszshistoo), Tun (Sedrena tunna), Khair (Akailia Kechu), Semal (Bombex, Malabaricum).

In the Terai region tropical vegetation consists of Bamboo, Sabai grassses, Narkat reeds, Tarmarisk, are important vegetation species. Sumpy depression and stegrent water bodies are over grown with water reeds, while reedy grasses, bulrushes and tamarask grow in the mergins. The high altitude of permanently frozen region and the permafrost region periglacial–glacier environment are predominant, where even green types of vegetation are found. Thus region is characterised by very thin and sparse vegetation. In the frost of Mahabharat range in the middle Sapt Kosi river basin, due to frontal characteristics of monsoon winds some evergreen type of vegetation are found; such as Oak (Kohodo dendendro) Birch, Pine, Devdar and Fir. The upper Kosi fan consists of moist deciduous and tropical moist broadly forest. Almost all Terai, upper Sapt Kosi fan and lower Sapt Kosi fan are well developed. In forested zone of great Himalaya bitch, Juniper, Blue pines, Bamboo and Rodo dendroa grow. In the high altitude areas vegetation are Dharf and Shurbs. On the basis of the regional variations in the climatic conditions vegetation type also varies and the vegetation of the Sapt Kosi river basin may broadly be grouped under three categories.

It is evident from the table (2.10) that Sapt Kosi river basin is characterised by a complex vegetation type due to the variations of the climatic conditions. Total forest cover of the basin consists of 88,731 sq. km which is 97% of the total area of the district. It is also evident that the subtropical Himalayan forest covers an area of 34,183 sq. km which is 37.4% of the total area of the basin. Mountain type of vegetation covers an area of 32,016 sq. km is under temperate zone where student and conical trees are found. The vegetation has a wide spread influence and landform impact in landform development and in the operation of geomorphic process. vegetation cover minimizes variation in soil flowed down as forest cover area receives relatively less amount of insolation. Vegetation influences transportation mechanism by different geomorphic processes. Dense vegetation cover minimizes over land flow maximizes infiltration of rain water. Conversely, open areas generate maximum land flow because in case of strong rainstorm with high rainfall intensity, rainfall water exceeds water absorption capacity of ground surface and hence instantaneous overland flow is generated and sub oil remains dry. Dense green covers lessen surface runoff and transportation and erosion by it more than forest cover. Considering the geomorphic significance of vegetation Tricart and Carlleaux have opinioned that 'Live climate vegetation type also produces its distinctive assemblage

Table - 2.10

BROAD VEGETATION TYPES FOUND IN DIFFERENT CLIMATIC ZONE

S.N.	Climatic Zone	Country (Terittory)	Vegetation Type	Area Covered (in sq. km.)	Percentage of Total Area (in %)
1.	Sub-Tropical (Am)	India-Nepal	Bamboo, Sabri Grass, Narkat reeds, Deciduous, Sal, Sagwan	56,715	62
2.	Temperate (H)	Nepal-China	Oak, Birch, Pine, Devdar and Fir etc.	32,016	35
3.	Freezed Zone (P)	Nepal-China	Summer flowers, Mosees and Lichens etc.	2,745	3
			Total	**91476**	**100**

Source: Computed from SOI Toposheet No. 72/G & 72K and Google Map.

of landform (Verstappen, 1954).

SOIL : Soil is the thin surface layer on the earth comprising natural particles form by break down of rocks, decayed organic materials, living organisms, water and air. It is formed under specific natural conditions and each of the elements of the natural environment constitutes to this natural complex process of pedogenesis. The soil formation is mostly related to the parent rock material surface relief. climate, natural vegetation, animals, insects and man also play an important role in the soil formation.

Sapt Kosi river basin is an extensive fluvial basin having heterogeneity basin, extensive valleys, rapid under cutting of the river channels and deposition of sediments to the local and regional base level. The whole higher Himalayan ranges, Tibet upland and an extensive area of Middle Himalaya, Shivalik ranges, extensive piedmont area (Terai) and highly dynamic plain tract.

CHAPTER – 3

GEOMORPHIC PROCESSES

Geomorphic processes lived their distinctive imprints upon landforms and each geomorphic process develops its own characteristics assemblage of landforms (Thornbury, 1954). Geomorphic processes are all those physical and chemical changes which affect a modification of earth superficial forms (Thornbury, 1969, p. 34.). A geomorphic agent and agency is a natural medium which is capable of securing transporting earth material (Thornbury, 1954). Landform development is that landform changes in orderly manner as processes generate through time such as under uniform external environmental conditions and orderly sequence of landform develops (R.C. Palmquist, 1975). 'Landscape remains in equilibrium conditions, their history is rhythmic punctuated oscillatory changes and their forms are punctuated by frictional rhythms arising out of the mechanism of driving and resisting forces. The identification and quantification of frictional rhythm (processes) and determination of their (of processes) dynamic competition is the major geomorphic problem and the main task before the geomorphologists is to solve this problem (Gilbert, 1909).

Interpretation of geomorphic processes and landform resulting their form has remained a long practice before the geomorphologist to go with detail investigation of the processes involve in the development of landforms. 'According to first law of Thermodynamics in any system of constant mass, energy is neither created nor destroyed but total energy remain constant and it can be transfer from one type to another type. While the second law of thermodynamics state that as time passes and the energy within the system becomes more equally distributed the entropy increases until the maximum entropy. All parts of the closed system have the same energy level (Chorley, 1985).'

In other words with the passes of time a system tends to achieve minimum energy and maximum entropy.

The geomorphologists involved in interpretation the landforms may be mentioned as process realism. G.K. Gilbert (1914), A.K. Sundborg (1956), R.E. Horton (1945), S.A. Schumm (1956). But this concept bloosmed with the work of A.E. Scheidegger (1961) and G.H. Durry (1972). It may be mentioned that a few geomorphologists engrossed with the process realism that they concentrated on the study of the mechanisms of physical and chemical weathering process at very micro-spatial and temporal scales requires a trained and expert geo-scientists in general and biochemist in particular, and secondly the results drawn through the investigation of processes at micro scales may not be applicable for the generalization of mechanism of processes at micro scale.

Geomorphic processes have the role in shaping the earth's configuration from the broader level to the micro level. Landforms at the broader as well as micro level are the result of geomorphic processes operating on them.

Problems of landform identifying is first and foremost responsibility of geomorphologists to have concrete idea who is in search of dealing with landform development of a particular area. Sapt Kosi river basin is very peculiar basin as its having a heterogeneous location where the landforms has been developed by a complex of geomorphic processes. The whole basin is one place and open system and at the other an enclosed entropy. Here, we have firstly gone through the general processes operating to develop landforms. All the processes involving the development of landform have been grouped under to major category i.e. the indogenetic processes which are involved in the development of landform within the earth crust whereas the processes involved in the development in the landform originating from the outside of the earth crust are the exogenetic processes.

A BROAD OUTLINE OF THE GEOMORPHIC PROCESSES OPERATING FOR THE LANDFORM DEVELOPMENT IN SAPT KOSI RIVER BASIN: Sapt Kosi river basin extended over an extensive area. The whole basin is a typical example of complex processes operating for the landform development. The upper catchment area of Sapt Kosi river basin particularly in the higher Himalayan zone and in the Chinese territory, the region is under the influence of glacier and periglacier where as the whole Nepal area is under the high altitudinal terrain. The lower most catchment area of the basin is under the tropical climate conditions and under the fluvial action. Thus, Sapt Kosi river basin, on the basis of the climatic conditions the dominants of landforms and the processes of landform development operating therewith may be grouped under the three broad categories.

A. Glaciated part of the basin

B. Periglaciated part of the basin

C. Fluviated region

The detail of the geomorphic processes region have been outlined in the table no. 3.1.

Table - 3.1

GEOMORPHIC PROCESSES ZONES

S.N.	Climatic Regions	Geomorphic Processes Zone	Area under Geomorphic Processes	Percentage % of Area
			Sq. km.	In %
1.	Polar (ES)	Glacial	2,835	3.1
2.	Highland plateaus and mountain (Temperate)	Periglacial	1285	1.4
3.	Tropical monsoon (Am)	Fluvial	87,356	95.5
		TOTAL	**91,476**	**100**

Source: Computed from IGIS software (version 1.1) with the help of produced map

It is evident from the table that the Kosi river basin undergoes a complex of geomorphic processes involve in the landform development in the basin. As it is stated that the complexities in the landform development is more common than simplicity (Thornbury, 1954).

Table - 3.2
GEOMORPHIC PROCESSES

Endogenetic Forces						Exogenetic Forces		
Diastrophic Forces						Sudden Forces		
Epierogenetic Forces		Orogenetic Forces				Volcanic	Earthquakes	1. River Water
Upward Movement (Emergence)	Downward Movement (Submergence)	Tensional Forces (Crustal Fracture)		Compressional Forces (Crustal Bending)		1. Fissure 2. Central		2. Underground Water 3. Ocean Water
		Cracking (Faults)	Faulting Up-warding	Warping Down-warping	Folding (Folds)			4. Wind 5. Glaciers

Source: Singh, S. 2006: Physical Geography, Prayag Pustak Bhawan, Allahabad, p. 82.

The following processes of landform development may be outlined which are involved in the basin.

1. Epigene or Exogenetic processes
2. Gradation
3. Weathering
4. Mass wasting or gravitative transfer
5. Erosion (including transportation) by running water, ground water, wind, glaciers
6. Aggradation by running water, ground water, wind, glacier

Epigene or Exogenetic processes: The forces shaping the forms of the earth originating from the atmosphere, they are also referred as denudational processes or the destructive forces. The present configuration of Sapt Kosi river basin is the outcome of the long exogenetic forces. These forces are continuously engaged in destructive of the relief feature acted by endogenetic forces through the weathering erosional, and depositional and therefore, planation processes. Exogenetic processes are active in the upper, middle and as well as lower catchment area of the Sapt Kosi river basin. The forms of the exogenetic processes involved in the landform development in Sapt Kosi river basin have been shown in the tabular form underneath.

Exogenetic and indogenetic processes are competitive forces which are engaged in continual conflict and the integration between these processes produce complex set of physical landscape. The exogenetic processes are gradational or planation processes involved in irregularities created by indogenetic processes through denudational mechanism and depositional activities. The planation includes the earth processes irregularities is completed through i) Degradational (level down), and ii) Aggradation (level up).

Sapt Kosi river basin is the typical example for all the geomorphic processes operating for the landform development in the basin. Generally, exogenetic processes accomplishes the erosional uncover through erosion, transportation and deposition also known as destruction processes.

The erosional work by different processes is performed through the mechanism of

i) Chemical erosion (corrosion or solution)
ii) Corrasion or abrasion
iii) Attrition
iv) Hydraulic action
v) Deflation
vi) Plugging
vii) Polishing
viii)Cryoturbation

It is hereby, important to mention that the gradation 'all those processes which tend to bring the surface of the lithosphere to a common level.'- Chamberlin and Salisbury, 1940, pp. 117-124. The gradational processes are involved in bringing the surface to a common level through weathering, mass wasting and erosion.

Weathering is the disintegration and decomposition of the earth material/rock in situ. These processes are really a process which acts collectively at and view of each surface and reduce solid rock masses to the clastic state. 'It is a static processes and doesn't involve the season and removal of material by a transporting agency' (Thornbury, 1954).

The large part of the Sapt Kosi river basin also undergoes through the weathering processes particularly the upper reaches and high altitudinal area of the basin is under the effect of low temperature and weathering condition. The high altitudinal area, owing to low temperature undergoes the weathering processes, alternate cooling and heating in the upper reaches has given high impetus to the weathering processes.

Mass wasting involves the bulk transfer of masses of the rock debris down slope of the direct influence of gravity (Thornbury, 1954). Mass wasting consists of landslide, debris avalanches, earth flows, mud flows, sheet flows and slope wash. 'Mass wasting is the usually added by the presence by water but the water is not such an, amount as to be consider as a transporting medium (Highway Research Board, Washington, 1958). The mass wasting is an effective degradational or denudational process in Sapt Kosi river basin. 75% of the area of the basin is under high altitudinal area ranging from 40 m to 8,848 m. consists of the Shivalik range, the middle Himalaya, higher Himalaya and the Tibet plateau. All the regions are under the direct influence of gravity consists of large scale mass wasting.

Mass wasting is an important plemtion owing through different processes which cause movement of land mass, from the ground surface along the hill slope and undulating terrain. The processes of weathering where disintegration and denudation of rock and the upper crust of the earth which are removed down slope by mass wasting apart by erosion by geomorphic agents. The nature of mass wasting can be grouped into different categories. The classification has been made by number of geomorphologists such as Sharpe (1938), Hutchinson (1969), Varnes (1958), and others. All of the taking account the type of movements and nature of the materials vary considerably ranging by chunks of rocks and the final partides of soil, in soil creep the unconsolidated, sands, silt, clay quantity move down slope as laminar flow without the formation of any channel where as causes material with massive movement of chunks of rocks, boulders mud and earth are removed by different processes as avalanches, landslide, mud flows and earth flow. The mass wasting in any terrain is evidently the formation of the rainfall conditions, slope of land, and the trace of earth related conditions the gravitational from acts for the downward movement of the mass initiated by heavy incident, frost-shattering and glaciated action in the glaciated region. The Kosi river basin experiencing dominantly the tropical rainfall has a high incident of mass wasting-mass wasting process during that time. Any sporadic rainfall of high intensity causes mass wasting. Sapt Kosi river basin consists of 75% into catchment area under the high altitudinal mountain range of Himalaya.

Mass wasting phenomena in the Himalayas are common, in the higher mountain area of the main Himalaya. The avalanches and rock fall due to glaciers are predominant where as on the lower periglacial zone the processes of frost shettering and solifluction are dominant. Rock fall due to disintegration of ricely blows by mechanical weathering process is also fairly widespread. In areas below periglaciated zone particularly in the Nepal region the rainfall causes hazard of earth flow, mud flow and landslides. Landslides are the most hazardous and recurring features in the habitated zone of lower Himalaya affected the roadway and human settlement to a great extend. Landslides are quite common on Himalayan slopes which have been responsible for the both permanent and temporary lakes (Pascoe, 1964).

Landslides are generally, occur along the Arun, Tamur and Sun Kosi in the Nepal Himalaya, A study of landslides conducted on the Arun basin by Basu and Ghatowar (Basu, & Ghatowar, 1988), thus all the hillslope of the Nepal Himalaya are subjected to mass wasting processes. The slopes of lesser Himalaya are generally fond to be convex in characteristics. Evidentially, indicating the action of the soil creep where as in the glaciated ridges, the crests and slopes are of the form of serrated ridges as concave 'U' shaped valley and hillslopes (Kaldova, 1972). Rill gully and reverine and bad line formations the geomorphic processes pressure completely through typical soil erosion and trending action of newly-establish feeble ephemerad channels in the unconsolidated valley slides and the upper hilly catchment or areas vulnerable to easy soil erosion have been cause of the formation of rill which with the progressive integration an intrachment of the valley leading to the formation of gullies and reverine. In Sapt Kosi river basin of Nepal Himalaya the channel are generally highly susceptible to erosion because of the incident of rainfall and ultimately the whole areas degenerate into a badland terrain composed of entracrced valley with steep side slope and scarps, unsuitable for agricultural processes. The processes of the soil erosion the sediment yield transported the bed into a dissected and undulating terrain.

Gradation: The three distinct degradational processes are weathering, mass wasting and erosion. These are all the gradational processes, in which the mass wasting has been already discussed above.

Weathering: Row weathering or weathering is the disintegration and decomposition of the earth material in situ. It is an important geomorphic process involves in owing the surface to a common level. 'It is mechanical fracturing or chemical decomposition of rack by natural agents at the surface of the earth (Sparks, 1960).' 'Generally, weathering refers to the breakdown or disintegration and decomposition of rocks in situ through mechanical effected by water, temperature, wind, different atmospheric gas and organism provided that there is no large scale transport of weathered product by denudational processes, except mass movement of rock waste, down the slope and the impact of gravity (Singh, 1998). Generally, weathering processes are conveniently divided into physical, chemical and biochemical processes, but there are so intimately interrelated that it is pretlical difficult to isolate one process to other. The wreathing agents are moisture, water, frost, insolation, wind. Chemical weathering agents are oxygen, carbon dioxide, and hydrogen, biological weathering agents are vegetation, animal and micro organisms. The physical weathering are commonly seen in the upper catchment area of Sapt Kosi river basin where as the chemical weathering are in operation throughout the basin, that biological weathering processes are is in mainly middle and lower Himalaya are specially in the Shivalik Himlaya.

Erosion/Fluvial processes and forms: The fluvial processes in the Kosi river basin are the most widespread over the entire basin except a small extent for the areas under the diversity of fluvial gemorpholgy demands a much more intensive research and wider coverage. The fluvial processes involves erosion which include mechanism of corrosion (dissolution of the soluble material through the product as of disintegration and decomposition of the carbonate rock), abrasion or corasion (removal of loosen materials of the rocks by different erosional processes in different manner). Abrasion is a strong geomorphic processes operating in the basin area and includes the breakdown of rocks and removal of loosen materials of rocks of valley walls and valley flows with the help of erosion tolls several landforms like pot holes, rapids etc. develops.

Hydraulic Action: This involves the breakdown of rocks due to pressure exerted by water currents of the rivers. This geomorphic process is highly violent in the whole upper and middle catchment area of Sapt Kosi river basin. Owing to high degree of slope and current in the water discharge in the river leads to vertical incision resulting into deep and narrow river valleys.

Attrition: It is a geomorphic process referred to as mechanical tear and wear of erosional tools suffers by themselves. This suffering in the boulders, cobbles, pebbles, while moving downing stream with water collide against each other and material involve in this process gets finer. It also acts as a tool in breaking the underline materials.

The fluvial processes (the erosional works) also involves in transporting the materials which are accomplish through floatation, suspension, traction, saltation and solution. 'G.K. Gilbert propounded a low of stream transportation based on the relationship between stream velocity and its transporting power (Gilbert, 1909).' The transported materials forms and extensive deposition but not involves in the development of any landforms, but the depositional work of the fluvial processes are very significant as it involves in the development of various types of landforms. Sapt Kosi river basin is a typical basin as the upper and middle catchment area under the devere geomorphic processes of fluvial action where as the lower catchment area for the depositional landforms. The deposition by the streams affected by verity of channel factors, such as:

i. Decrease in channel gradient
ii. Spreading in channel flow
iii. Obstacle in channel flow
iv. Decrease in volume and discharge in water
v. Decrease in stream velocity
vi. Increase in sediment bad

Decreases in stream velocity reduce transporting power of stream which is forced to additional sediments load to settle down. The landforms develops due to the deposition of material are commonly observable in the Sapt Kosi river basin in Indian territory, extensive flood plain, Khadar, bhangar, landforms and the migration behavior of the stream leading to the development of various landforms.

CHAPTER – 4

DRAINAGE DEVELOPEMNT

'Drainage refers to the manner in which precipitation falling with in an area and brought from outside is drained off (Ahmad, 1965).' It comes under the domain of Fluvial Geomorphology and in real sense and involves in the study of river dynamics including runoff, discharge, erosion, transport of suspended sediment load and bed load, mechanism of sedimentation, river morphosis, drainage types, drainage patterns, channel forms, and patterns and Limits to drainage development and the resultant landforms etc. On the basis of the primary data derived through field measurement and mapping, instrumentation and laboratory experiments and the analysis, on the parameters (K.R. Dixit), has rightly pointed out that 'Fluvial geomorphology as different aspect has not received the attention it deserves from the geomorphologist. Infect, they are did not simply exist a tradition of fluvial geomorphology in India till Leopold's fluvial processes in geomorphology appeared on the scene and give a new orientation to the discipline (K.R. Dixit, 1983, p. 5). Most of the works on the fluvial geomorphology are related to the evolution and development of drainage and drainage pattern calculation and depiction of longitudinal and transverse profiles on the basis of topographical maps, calculation of sinuosity indices and mean meander properties and qualitative description of fluvially originated landforms. Recently field measurement and instrumentation in fluvial processes have been initiated (Sarvindra Singh and Alok Dubey, 1995; V.S. Kale, D.C. Goswami, Gowahati and Almora, J.S. Rawat etc.). Several other authors are based on the significant contribution in evolution and development of river basins.

Significant contribution in the evolution and development of the drainage basin, drainage patterns have been made by

H.L. Chibberm (1954): drainage pattern observed in India and adjacent countries 1954; Thrust superimposed drainage of the Shali area (K. Bagachi, 1960); Drainage pattern in Ganga delta, B. Ghosh and S. Singh, 1963; Reconstruted drainage system in Luni basin; A.B. Mukherjee, 1970; Terminal Trangular trackts of island streams in Haryana, A.K. Pal (1970); Analysis of Belan river, West Bengal, S.R. Basu and S.C. Chakraborty 1972; Some consideration on the decay of the Bhagirathi drainage system, K.M. Hironi, 1982; Evolution of the antecedent pattern: A study of Banas, A.B. Mukherjee, 1993; Morphology and drainage evaluation of Sarswati basin.

Enormous work has been done in the field of drainage evolution / Fluvial Geomorphology by Indian Geomorphologists. A critical review of the status of Indian rivers carried out by V.S. Kale (2002), Harizon et. al. (2004), Jain and Tondon (2003), Jain and Sinha (2004), V.S. Kale (2003), Kale and Barakar (2006), Kale and Joshi (2004), Sarmha (2004), R. Sinha (2005), M. Jale (2003), K.S. Valdia (2002).

In the light of the works of the above eminent geomorphologists the drainage evolution of the Kosi river basin has been evaluated. 'A drainage basin, alternately described as a catchment area, is in Geomorphology and hydrology a region drained by a particular stream or by a river system. The amount of water reaching the stream is dependent upon the size of the basin, the total precipitation and the losses due to evaporation and to absorption by soils and vegetation (Egyed, 1957).' the drainage system and evolution may be analysed with respect to time: Perennial Permanent flow, seasonal intermittent, periodical, episodical and ephemeral. The drainage also has a pattern: dendratic, trellis, crow-foot, rectangular, pinnet, radial and angular. It is also evaluated in terms of direction type etc.

The evolution of drainage network of a particular region is approached in two ways: i) Descriptive approach, ii) Genetic approach (Fairbridge, 1968). Descriptive approach involves forms and pattern of a stream where as genetic approach involves the investigation of the evolution of streams in relation of streams, lithology and structure. Thus, the drainage system of a particular river basin is being evaluated.

The drainage system refers to the origin and development of streams through times while drainage pattern means spatial arrangements of form of drainage system in terms of geometrical shapes in the area of different rock types, geological structure, climatic conditions and the denudational history, for example dendritic and parallel etc.

The evolution of drainage basin is largely evaluated by two main factors i) Nature of initial surface and slopes, ii) Geological structure and the drainage system are divided in two broad categories the on basis of the adjustment of the streams to the initial surface and geological structure as a) Sequent streams: such as consequent streams, subsequent streams, obsequent stream, resequent streams. b) Consequent streams: such as insequent streams, this includes antecedent stream and superimposed streams.

The Kosi river which is major tributary of Ganga and forms the lifeline of the state of Bihar in north India originates at an elevation of more than 8,000 m in the Himalaya and finally discharges into the Ganga at Kursela. The Kosi river is also called Sapt Kosi in Nepal. It has seven tributaries namely the Sun Kosi, Arun, Tamur and in the upper reaches Kamla-Balan, Trijuga and Bagmati in the plain. Sapt Kosi river has total distance of 468 km passing through Tibet, Nepal and India. Sapt Kosi is known for shifting latterly and thus creating the problems of flooding causing loss of human life. The major tributaries of Sapt Kosi river joining the main river are Triguja, The Kamla-Balan, Bhutahi Balan and Bagmati. All the tributaries are joining from the right in the plains of Bihar. The distribution of the catchment area has been shown in table no. 4.1.

Table - 4.1
DISTRIBUTION OF CATCHMENT AREA IN SAPT KOSI RIVER BASIN

RIVERS	IN INDIA	OUTSIDE INDIA	TOTAL
	km^2	km^2	km^2
Kosi including hilly tributaries	11,070	63,430	74,500
Kamla Balan	2,980	2,465	5,445

Bagmati	6,320	7,080	13,400
Trijuga	-	706	706
Bhutai Balan	-	1,105	1,105
Total	**20,370**	**74,786**	**95,156**

Source: Garde, R.J., 2006: River Morphology, p. 391.

The hilly tributaries are the Arun, Sun Kosi and Tamur. The Arun Kosi's the largest of the hilly tributaries which drains the mt. Everest. Its catchment area is 34,650 sq. km constituting and Contributes 30% of the flow 36% of sediment load, the Kosi at Tribeni. The Sun Kosi is the second largest tributary 19,000 sq. km and constitutes 44% of the flow and 42% sediment load at Tribeni. Tamur drains mt. Kanchenzonga. Its catchment area is 5,900 sq. km and contributes 19% of flow and 22% sediment load of the Kosi at Tribeni.

Sapt Kosi river is a river that has been originated with the start of from the beginning of Himalayan orogeny. Thus, it is an example of sequent streams. The sequent system of the Sapt Kosi river basin has been studied hereby in detail.

MAJOR DRAINAGE SYSTEM WITH THEIR GEOPHYSICAL PROPERTIES:

THE CONSEQUENT STREAM: The consequent streams follow the regional slopes. These are also called as deep streams. Sapt Kosi river with its all tributaries which gets their origin from the different ranges which are mostly a folded structure. The whole Himalayan belt is the result of intense folding (when the crustal rocks are folded due to lateral compressive forces into parallel anticline and syncline). The consequent streams of the Sapt Kosi river basin are formed by the synclinal troughs. That is why these streams are also called as synclinal consequent streams. Sapt Kosi river as a master stream is a best example of sequent stream because the evolution of the whole Sapt Kosi river drainage basin has taken place after the evolution of different ranges of the Himalaya. The channels of the basin have taken their origin after the establishment of the regional slope. Most of the streams draining to the master stream are also sequent streams. The consequent streams are divided into two types. **a)** Longitudinal consequent and **b)** Lateral consequent.

Lateral consequent are all the tributaries of Sapt Kosi master stream. These are Indrawati, Bhote Kosi, Tama Kosi, Sun Kosi and Tamur. All the streams joins the master consequent stream of Kosi at different places more or less right angle.

Sapt Kosi river which is the master consequent river follows the regional slope of the area. On the eastern side is the ridge separating it from the Mahananda catchment area and the west is the ridge separating from the Gandak (Budhi Gandak) catchment area. There is 87 m. drop in elevation in the 160 km reach between the Chatra gorge and Kursela near the confluence in Ganga. Nearly 80% of total catchment area of Sapt Kosi lies in Tibet and Nepal consist of 74,780 sq. km. these areas having the general regional slope towards east and south-east. The slope of the catchment area varies from 32 to 90^0. The three hilly tributaries are Arun, Sun Kosi and Tamur. The Arun is the longest of the hilly tributary which drains the Mt. Everest and has got its origin after the established slope. Its catchment area is 34,650 km sq. and it contributes 37% of the flow. It also follows the regional slope of the area. Most of these tributaries are Kamla Balan, Trijuga and Bhutahi Balan.

SUBSEQUENT STREAMS: The subsequent streams originate after the formation of the master consequent stream. The subsequent stream always follows the axis of the anticlines or ridges and the strike of the beds. Some geomorphologists are in opinion that the subsequent streams originates on the flanks of the anticlines and join the master consequent at almost right angle. 'It should be noted that all the first generation of tributaries to consequent streams over subsequent in the severe that they arise subsequently to the establishment and incision of the consequent streams (Wooldridge and Morgan, 1960).' 'such streams starting as gullies on the side of the primary subsequent stream discover and explore bed of structural weakness due to softer strata, fault or joint planes and shatter zones' (Wooldridge, S.N., pp. 173-174). 'The term subsequent may be used generally note to indicate choronological sequence but a relation between drainage on the one hand and geology, structure and tectonics on the other (Ahmad, 1985).' Generally, subsequent streams refers the streams traverse to the master consequent. Men Chu, Nien Chu, Kamla Balan, Bagmati, Trijuga, Bum Chu, Phung

Chu, Tama Kosi, Bhote Kosi, Indrawati tributaries of the Sun Kosi river are the examples of the subsequent streams.

OBSEQUENT STREAMS: The streams flowing of opposite direction of the master stream are called the obsequent stream (Singh, 2006; Reprint). Obsequent steams are also consequent because they also follow the micro level slopes of the ranges. The streams originating from the interfluvial track of the same river basin are the obsequent streams. The northward flowing streams join the subsequent stream of Sun Kosi and run west to east as obsequent stream because they are opposed to the direction of master consequent like the Kosi, most of the stream northern slopes of the Shivalik streams of the southward flow master stream of the Kosi are the obsequent streams. Yaru Tsangpo originates near almost north of Kanchanzonga mountain ridge (north eastern corner of the basin), and flows towards northward after 189 km running it meets with Arun river.

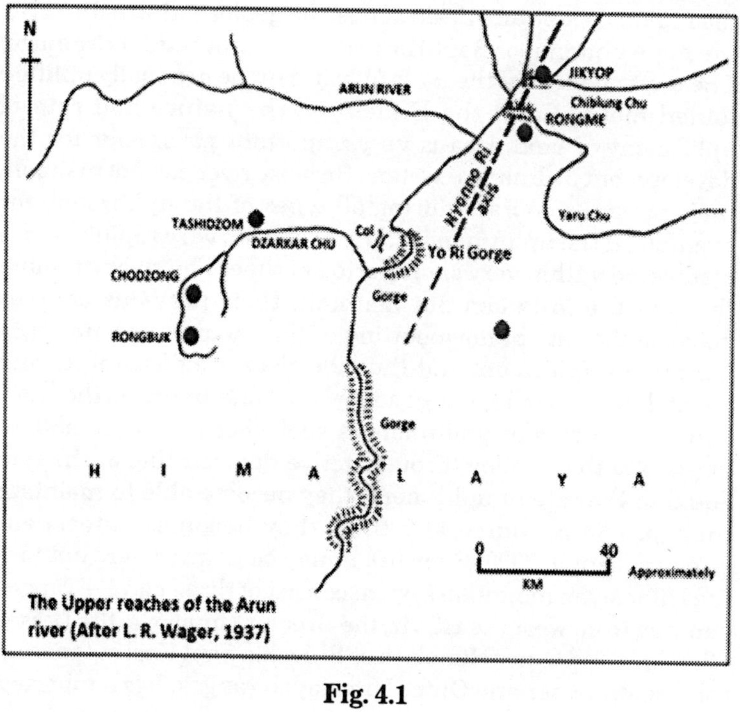

The Upper reaches of the Arun river (After L. R. Wager, 1937)

Fig. 4.1

RESEQUENT STREAMS: The streams flowing in the direction of master consequent stream refers as the resequent. It is an accepted fact that during the initiation of 2nd cycle of erosion in the folded structure are the resequent streams. The geological denudation of folded mountain during the 1st cycle of erosion results in to inversion of relief with the passage of time. Thus, the stream develops in the synclinal portion become resequent streams which are through analogous to the original longitudinal consequent develop during the 1st cycle of erosion are the resequent streams. For example Khimti Khola, Belephi Khola, Malemchi Khola, Dhankuta Khola etc. are the resequent streams which are the parallel to the master consequent streams.

INSEQUENT DRAINAGE SYSTEM: The development and evolution of Sapt Kosi drainage basin has been traced out as insequent drainage system. Satellite imageries and field observations revels the fact that the main master channel of Sapt Kosi river is a insequent drainage system, it do not follow the regional slope and drain across the geological structure. So, the main channel of Sapt Kosi river is a antecedent drainage. The upper course of the river fully lies in the gradually uplifted folded mountain of the Himalaya. The nature and rate of upliftment of land area is very important parameter for the development of drainage system. Because river can not maintain their previous courses during all types of the upliftment, for example, if the upliftment of an area occurs very rapidly and is completed within very short period of time. The river draining through the area can not maintain their previous courses because the rate of downcutting of their valleys can not pace the rate of upliftment and thus the river are dismember and adopt different courses depending land conditions. On the other hand, if the rate of upliftment is such that rivers are able to degradate their valley through active downcutting at the rate equal to the rate of upliftment, they become able to maintain their previous course and thus, they becomes antecedent streams (Singh, 2006; Reprint). It may be pointed here out that the Himalayan mountain system consist of three parallel ranges running from west to east, viz, the Greater Himalaya, the Lesser Himalaya and Outer Himalaya or Shivalik. Sapt Kosi cuts across the Great, Lesser and Outer Himalayan ranges. It is supported by the facts of paleochannels that the river was present before the folding and upliftement of Himalayan ranges. With the upliftment of these ranges of Tertiary period the Sapt Kosi river has maintained southward flow through downcutting their valley in harmony with the rate of cutting. Sapt Kosi river has formed deep and narrow gorges while cutting the Himalayan ranges. Fig no. 4.1 clearly depicts that the Kosi river has maintained their courses during the successive upliftment of Himalaya. The Arun river has been quated as the best example of antecedent river by L.K. Waiger (1937). The Arun river flows from west to east in its upper reaches following the structural grain of the region to the north of Himalaya. It's suddenly turned due southward and drains across the Himalaya through several steeply entranced narrow gorges and thus, presws to be an antecedent drainage.

DEVELOPMENT OF DRAINAGE SYSTEM IN KOSI RIVER SYSTEM: From the above concept of evolutions of drainage basins, the master stream of Kosi river as well as its main tributaries have been analysed herewith keeping in view of their evolution.

YO GI GORGE: A few geologists opine that the sudden southward bend of the Arun river is due to river capture. The advocates of the antecedent origin of the Arun river maintain that the convex course of the river across the Himalayas could be possible only through active down cutting of the valley in accordance with the rate of upliftment. A series of terraces on either side of the valley of the Arun river also validate its antecedent origin.

Wager does two things: first he describes the Arun from its headwaters to its journey through the Himalayas. Secondly he gives discusses possible antecedence of the Arun, sitting the Yo Ri Gorge as evidence. It should be noted, that antecedence of the major river systems in the Himalaya is still unresolved.

The upper Arun, that runs parallel to the range, is a wide, braided river, low gradient, cutting through a wide valley of alluvial fill (mostly gravel). The lower Tibetan Arun is more incised, single channel river. In many areas the valley is still wide, but the river has cut into the gravel, often leaving terraces behind. The distance between the upper terraces and the channel increases downstream. It is along this stretch of the Arun that the Yo Ri gorge exists.

As the river gets to the Himalaya it enters the main gorge. The slopes are very steep throughout the decent and the river cuts deeper and deeper into the mountains forming fantastic gorges. The entrance to the main gorge can be seen above. The flat fairly wide valley gives way to the deep gorge and high slopes surrounding the river. This continues until the river exits the range, and the slopes shallow out.

In addition to geomorphologically characterizing the Arun, Wager offered some evidence for antecedence. His main piece of evidence is the Yo Ri gorge which I have mentioned a few times already in this talk. Just to remind you, the Yo Ri is just south of where the river makes the abrupt bend and becomes incised in the alluvial gravels. The next picture will be a blow up of the area within the yellow square. The Yo Ri Gorge, which you will see in a moment is a magnificent gorge with over 6,000 ft. of relief.

Describe how the river in the upper portion of the picture is a single channel moderately incised river cutting into alluvial gravels that fill a wide flat valley. When it enters the Yo Ri, it abruptly turns out of the valley, and cuts a deep gorge into the harder gneisses of the surrounding range. After making another sharp turn, the river returns to the valley and continues cutting into the gravels again.

Opposite view, showing the entrance to the Yo Ri, the river would be predicted to continue to cut through the gravels within the valley, but it takes a sharp turn and cuts through less easily erodable rocks instead.

KOSI CHANNEL IN MEGA FAN AREA: There are two theories existing regarding the original course of the Kosi viz.

i) That it joined the Brahamputra and

ii) That its former channel was the one now occupied by the Karatoya (a river in Bengal).

The second theory was put forwarded by Mr. Donnell in the

Bengal census report, of 1891. He regarded. "The Karatoya in the days of its real greatness was the bed not only of the Teesta but of the Kosi and of the numerous hill streams which now unite to form the Mahananda".

The first theory which was advocated by Dr. B. Hamilton has insufficient evidence in support of the suggestion that the kosi joined the Brahamputra. However, Mr. Shillingford wrote: "The Kosi in remote times broke away eastward in part along the present bed of the Parman or Panar" which is in divergence from the theory advanced by Dr. Hamilton.

Although no historical records are available, yet it does not seem improbable that the Kosi joined the Mahanada through the present course of the river. Parman near Araria in Purnea district in some remote past. The local belief is that in some by gone days, the Kosi used to flow along the course of the Parman. There is a long stretch of depression varying in width from 30 to 60 m, passing from Forbesganj side towards Araria and further south, which the local people believe to be the one time course of the Kosi. The different past course of the Kosi for which records are available are discussed below.

THE KALI OR KARI KOSI: It was known in its upper reaches as the kamala and in nepal as the kajri. It flowed about 1.5 km.

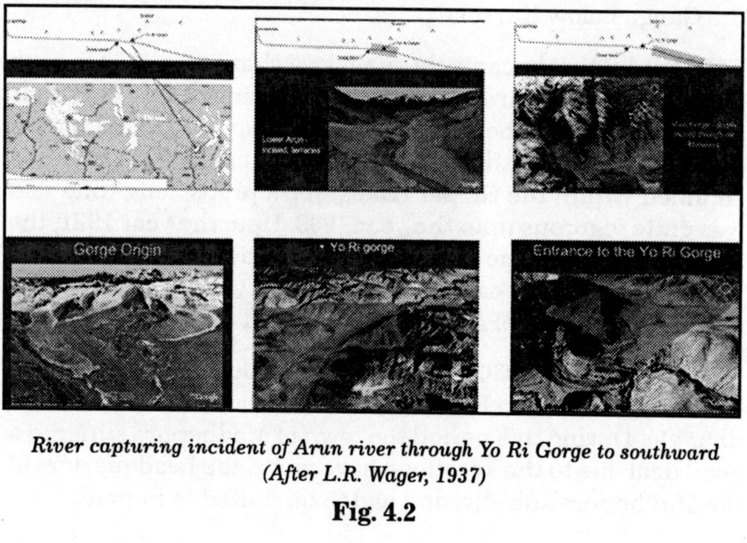

River capturing incident of Arun river through Yo Ri Gorge to southward (After L.R. Wager, 1937)

Fig. 4.2

west of the civil station of Purnea. The name 'Kali' as given to this river is from the curious fact that this is the only abandoned channels of the kosi, which carried dark limpid water. This was the main kosi in 1731, forming at that time the western boundary of the Purnea district.

THE DHAMDAHA KOSI: It is known in the higher reaches as the Pheriani. This was shown as the main kosi by Dr. B. Hamilton in 1807-11. Between the Kali and the Dhamdaha kosi, there is another silted up bed, called in upper reaches the Livari and in lower down reaches the Burhandi, which must have been occupied by the main kosi between the year 1731-1807, but no record in this regard is available. This channel crosses the North-Eastern railway line between Karhagola Road and Kursela Railway stations, and joins the Ganga near Karhagola.

THE HIRAM: It was the main kosi as per the Revenue Survey maps of 1840-47, flowed to the west of the Dhamdaha thana and about the year 1870 began to throw off branches into the Daus Swamps, which became the main kosi in 1873 and flowed into the Ganges opposite Patharghato.

THE DAUS: It was the main kosi from 1873 to 1893 which marked the further westward swing of the kosi. The Daus flowed into the Ghugri (coming from west) and they conjointly felt into the Ganga below Kursela.

THE LOARN: It became the main kosi channel in 1893 though the kosi began to throw offshoots into it since 1891. The Daus and the Loran in their upper reaches used to be fed by the Sursar Dhar which still exists. The main kosi stream remained confined within the Sursar (Daus) for a pretty long time and was quite vigorous upto the year 1909. Upto the year 1921, the main kosi followed the course of the Mirchaia dhar which flowed about 13 km. to the east of Madhepura town during 1917, the straight kosi channel got completely silted up.

From 1922 to 1936 Sapt Kosi flowed through a number of channels such as the Belhi dhar, the Parwane dhar, the Tilawe dhar etc. During this period the town of Madhepura suffered a good deal due to the kosi flood havocs and the headquarters of the Madhepura sub-division had to be shifted to Supaul.

With regard to shifting of the Kosi River and its various channels, Bir bund deserves special mention. This bund was about 700 years old and practically continuous from a point about 5 km. south west of Bhimnagar and terminating quite close to the Soranga Range. The old line of borrow pits on the western side of bund became one of the main feeders of the Dhemra channels, when the bund used to get breached from time to time. The Tilawe depression also started from the line of borrow pits from the western side of the bund. There were many places between Bhimnagar and Saharsa where the Kosi used to spill into the Tilawe.

During the year 1924-26 a small channel was formed about 8 km south east of Bhaptiahi. This was an off-shoot from the Dhemra and became known as Udha channel. In the following years this channel gradually opened out. From 1928 to 1932 this channel was further enlarged. In the year 1934 the great earthquake of Bihar took place with its epicenter in the kosi catchment which caused disastrous land slides and enormous local subsidence in the catchments. In the next year i.e. 1935 there were heavy floods in the Sapt Kosi River. There was a big jump of about 13 kms. Towards west between 1930-32 and a bigger jump of 38.5 km. between 1936-38. The Udha channel then became the main channel of the kosi. The quick shifting between 1936-38 was a parently due to the quick silting of the old channels on account of the heavy silt brought down by the river as a result of disintegration of the rocks in the catchment due to the 1934 earthquake.

Nirmali was considered to be free from kosi floods upto 1938 but was affected by floods in Subsequent years. This village for many years had been protected from flood by various bunds and consequently had received no silt from Balan river which had built up the surrounding country by as much as 2.7 m Nirmali, therefore became a natural depression with the result that the kosi rushed to the low lying area and became active there. After 1939 the westward movement of the kosi was slow.

In the year 1936 the river Mahuli became one of the kosi channels. It received a fair quantity of the kosi flood discharge through a gap in the Bir bundh, In Nepal territory and carried the same down to the Beti below Raharia railway station. By this time the Dhemra was also carrying down a percentage of Sapt Kosi right bank spill through the gaps in the Bir bundh. Bhaptiahi also came under threat by the floods of a branch of the Dhemra, locally called the Udha dhar.

In the flood of 1936 the gap in the Bir bundh near Babia ghat (just to the north of Bhimnagar, in the then Bhagalpur district) was widened and in 1937 practically the whole of Sapt Kosi discharge was thrown into the Dhemra basin though it was carried mainly by the Dhemra, the Beti, the Tiljuga and a number of spill channels of the Dhemra. It was in this flood that all the culverts and bridges of the railway line between Nirmali and Bhaptiahi and Supaul were severely affected and most of them completely damaged. Since 1936 till the river was confined between the two flood embankments the kosi flood-discharge used to carried down by the Dhemra, the Tiljuga and innumerable right bank spills of the Dhemra between Bhaptiahi and Parsarma railway stations.

DRAINAGE PATTERN DEVELOPED IN THE KOSI RIVER BASIN: Drainage pattern refers to spatial relationship among the streams or river, which may be influenced in their erosion by inequalities of shape, rock resistance, structure and geologic history of a region (Fairbridge, 1968). It is a form of the drainage system which develops during the various stages of drainage system. As soon as the relief in any area determine or otherwise exposed above the sea level or after glacial retreat, surface wash from rains will start to produce valleys cutting into a new landscape. The stream begins to flow in the direction of maximum slope and produce the first ridges in valleys. Since water erosion occurs in all continental glacial areas, it has been found to be most convenient if we describe the direction of the landscape. The pattern of ridges and valleys in terms of the drainage pattern, several types of drainage pattern develops. It is an establish fact that through the drainage pattern of same region be similar but not the same, but there are some common characteristics which enables to distinguish different drainage patterns. Most of the drainage patterns are the result of either adjustment to the structure or lack to the adjustment of structure. Most of the patterns develops rivers of the Himalayan region are the origin of lack of the adjustment to the structure, because the drainage patterns which are the result of the adjustment of structure are found mostly in the stable landmass of hard rocks but it is inverse in case of a stable areas, like the Himalayas.

Zernitz (Zernitz, 1932) summerised seven types of drainage patter: Dendritic, Trellis, Rectangular, Radial, Angular, Parallel and Irregular. It is completely non-genetic classification though the drainage patterns of some regions may be similar but not the same. Generally, ten types of patterns have been identified. Sapt Kosi river basin marked by a region of superimposition is a common feature in tertiary mountain areas. It is due to the cause lack of adjustment to the structure. The following drainage patterns have been identified in Sapt Kosi river basin.

DENDRITIC DRAINAGE PATTERN: The term used by I.C. Russell (1898) and later vividly described by Cleland (1916). The dendritic or tree shaped drainage pattern is common in the Himalayan region. The dendritic pattern is associated with the homozonous lithologies, horizontal or very gently deeping strata

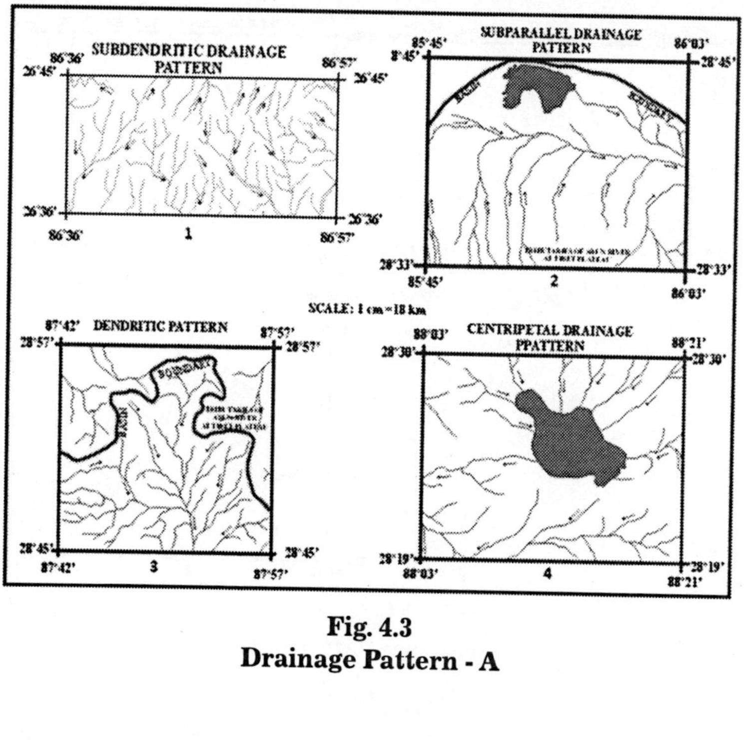

Fig. 4.3
Drainage Pattern - A

flat and rolling extensive topographic surface having extremely low relief. This pattern also develops variety of different structure and environment such as in the mountainous and hilly areas. The modified dendritic drainage may be described as pinnate, sub parallel or anastometic. The pinnate pattern is feather-like with tributaries parallel and joining the master stream at acute angles found at the junction of master stream and tributaries are so small as to make the major and subsidiary stream near the parallel. Most of the streams like Bhote Kosi, Indrawaty, Tama Kosi, Tamur and Arun's upper reaches and some other streams normally forms the dendritic pattern of drainage which is shown in Fig. no. 4.3(3).

TRELLIS DRAINAGE PATTERN: A pattern formed by the network of tributaries and master consequent streams which follows the regional slope and well adjusted to the geological structure. It is mostly developed in folded and tilted strata. Trellis pattern also developed glaciated area in the presence of linear glacier features as well as in the regions where parallel

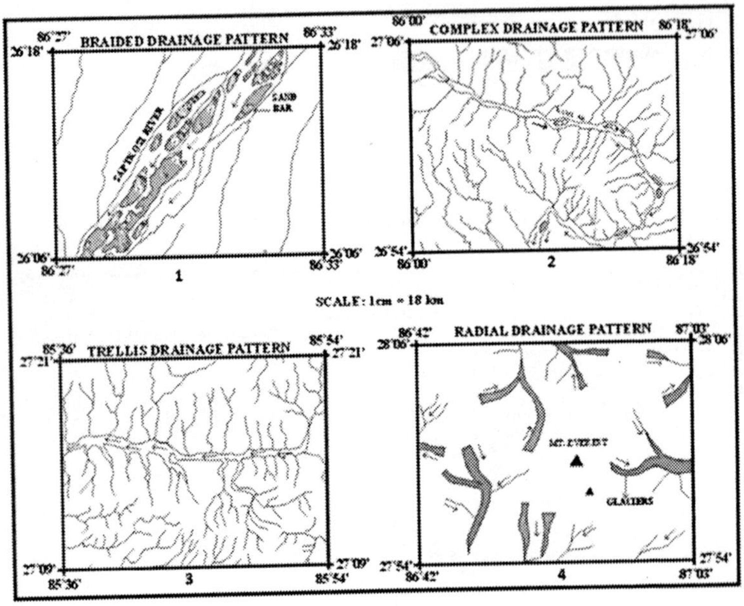

Fig. 4.4
Drainage Pattern - B

rocks of various resistance outcrops are found. Thus, the network of numerous longitudinal streams and traverse or lateral consequent is called trellis pattern. This pattern are commonly identified and most of the tributaries streams of the upper Kosi river basin as well as the streams that flows parallel in between the ridges of the higher Himalaya, middle Himalaya and the Shivaliks. The identified locations confirming trellis pattern in the Kosi river basin namely in Bum chu river, initial reaches of Kamla river, Shivalik reaches of Bagmati river, and some other region too. Example of trellis pattern is shown by Fig. no. 4.4(3) which is part of Shivalik reaches of Bagmati river.

RADIAL DRAINAGE PATTERN: Radial drainage patterns are develops on a structural or geomorphic quaquaversal basement, such as volcanic, plutonic, laccolithic dome, salt dome or circular monadonocks (Fairbridge, 1968). Although, Kosi river basin has a wide and extensive regime as it has been mentioned above that 62% area of the basin is highly undulating and mountainous region. The whole Himalayan region is a structural region characterised by folded, uplifted structural basins and conical shaped hills. From the analysis of the topographical map

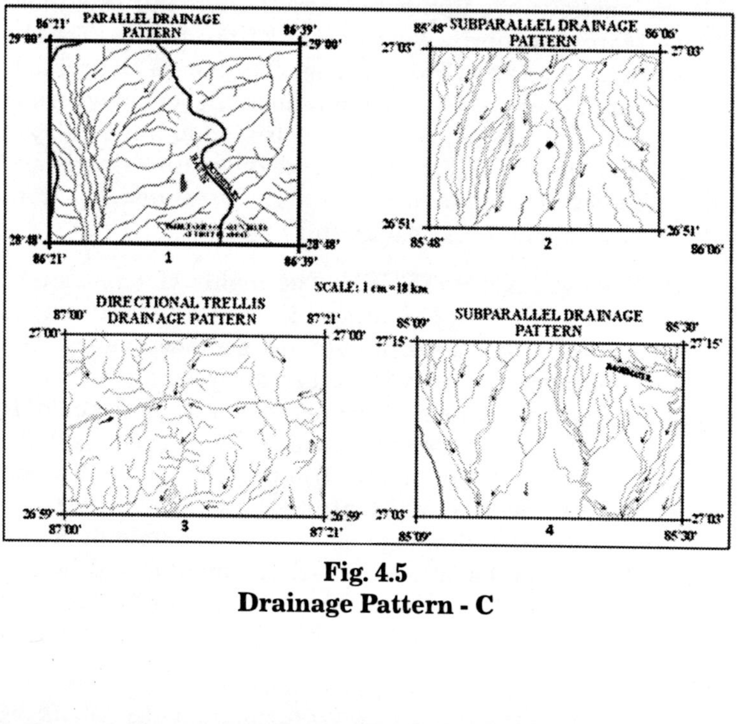

Fig. 4.5
Drainage Pattern - C

covering the basin, extensive field work as well as the Google maps we have identified the centrifugal as well as centripetal/ radial drainage pattern. The identified locations of such drainage patterns are mt. Everest, Gyachang Kang, Mt. Makalu, mt. Labuji Kang and some other mountain peak too; these mountain peaks is also a water divide line between Great Himalaya and Brahmputra (Tsanpo). This type of radial drainage pattern is shown by Fig. no. 4.4(4). This is a part of Khimti Khola and Likhu river drainage area in middle Himalaya in eastern Nepal.

PARALLEL DRAINAGE PATTERN: In the parallel pattern, as the name implies, consist of parallel master and tributary streams pronounced regional slopes, parallel folds and parallel topographic features. It is found in the topographic features, such as drumlins and lateral moraines which are conducive to the formation of this drainage type. This type of drainage pattern is mostly found in the parallel ranges of the Himalayan mountain under Sapt Kosi river regime. In the high altitudinal area drumlins and lateral moraines has been develops by the glaciers, which have formed parallel drainage pattern such drainage pattern locations are upper reaches of Nieh chu (Southern Tibet), Men chu (north-west corner of the basin), Ra chu (originate from north of Pasang Lhau chuli) these all drainage patterns are develops in glaciated influenced region where as some parallel drainage patterns are also found in Himalayan Terai region, such as Terai reaches of Bagmati, Trijuga, and some other small tributaries of Kosi river. The drainage pattern is shown by Fig. no. 4.5(1), 4.5(2), and 4.5(4).

HIGHLY VIOLENT PATTERN: The highly violent pattern shown in fig. no. 4.5(2) [Complex drainage pattern] is characterised by the areas of complex geomorphology. The complex drainage patterns in nature are result of differing lithology, regional slopes, presence of joints and geologic activities such as glaciations, limestone solution (Zernitz, 1932, Howard, 1967 and Thornbury, 1969) have given full description of commonly occurring drainage pattern and their interpretations, such drainage patterns are commonly found in Sapt Kosi river basin particularly in middle and higher Himalayan areas of the basin. This pattern is the result of successive upliftment of the Himalaya and high altitudinal glaciers as well as high degree of slopes. The middle and Higher Himalayan areas of the basin are the area of high slope which has given the development of the complex drainage patterns. These are also highly violent because of the high slope currents in the streams are also very high.

RIVER CAPTURE: The diversion of the part of the course of river by the river is referred as stream captures a stream capture/stream piracy/stream diversion or river capture. This situation in a fluvial region is commonly found in the region of high headward erosion by the valley lengthening and the lateral erosion.

It has attained the present shape after a long process of sub-areal erosion. 'A group of Geologists and Geomorphologists opined that the present drainage system of the Himalaya is the outcome of progressive river piracy during various stage of drainage development. The Arun river, a lateral tributary of the Kosi river has capture the Phung Chu. Such cases are common at the micro level too. This is due to the headward erosion (Mittal, Prakash and Bajpai, 1974).' The powerful stream pushes the water divide backward toward the side of gentle slope through active headward erosion. Prolonged headward erosion by move powerful stream flowing the steeper hillside of the divide result in the coalesces of the sources of both the stream on both side of divides. Since the valley flow of the stream of the steeper side of the divide lower the valley floor of the stream gentle side of the divide. Enhance, the former captures the headwater of the later.

CHAPTER – 5

MORPHOMETRIC ANALYSIS

The term morphometry, as define by Clark (1966) 'is the measurement and mathematical analysis of the earth's surface and its shape and dimension of its landform. The application of principles of mathematical and statistics to quantitative geomorphology is essential for the meaningful analysis and objectives of research are to be achieved. Morhometry includes not only the varerer characteristics of the river and drainage basin but also such aspects the area, altitude, volume, texture of the land as well as well as slope. Measurement of shape or geometry of any natural form- be it plant, animal or relief feature is termed morphometry (Strahler, 1969). Infect morphometry in corporate quantitative study of the area, altitude, volume, slope, profile of the bed and drainage basin characteristics of the area concern (Singh, S., 1985). The morphometrc analysis dealt with two distinct branches a) Relief morphometry, and b) Fluvial morphometry

Releif morphometry deal with the analysis of terrain characteristics though hypometric curves, clinographic curve, area height curves, altimetric frequency, histogram and curves, superimposed projected and composite profiles, which assists in dealing with different aspects of drainage basin.

Under the impetus supplied by Horton (1945), the description of drainage basins and channel network was transformed from a purely qualitative and detective study to a rigorous quantitative science capable of providing hydrologists with numerical data of practical value. The Horton work was developed by Strahler (1950, 1952, 1956, 1958) and his associates Melton (1957), Morisawa (1959), Schumm (1956). The linear aspect deals with the hierarchical order of stream, number and length of stream segment and various relationships among them and related morphometric laws such as laws of stream numbers and stream lengths. The areal aspects includes the analysis of basin perimeters, basin area and related morphometric laws, laws of basin area and low of alometric growth; stream frequency, drainage density and drainage texture. All three forms a part of fluvial morphometry.

Significant contribution has been made by different geomorphologists and in dealing with the basin morphometry from time to time. e.g. De Martonee (1934), P.S. Javanovic (1940), C.P. Peguy (1942, 1947 and 1948), R. de Smet (1951 and 1954), A.N. Strahler (1950, 1952, 1954, 1958, 1969), P. Birot (1955), H. bauling (1939, 1956, 1957, 1959), R.E. Horton (1932, 1945), J. Hanson-Lowe (1935), J. I. Clarke and K. Orrell (1958), M.A. Melton (1957, 1958, 1959), S. A. Schumm (1956, 1963), J.C. Maxwell (1955, 1960), H.W. Anderson (1957), D.R. Coates (1958), A.J. Brosco (1959), M. Morisawa (1959), C.W. Carlston (1963, 1965), L.M. Brush (1961), R.J. Chorley (1957, 1962, 1966, 1969, 1972), L.B. Leopold and W.B. Langbein (1962), K.L. Bowden and J.R. Wallis (1964, 1966), R.B. Boyce and W. A.V. Clark (1964), A.E. Scheidegger (1965), W.C. Auckermann (1966), K.J. Gregory (1968), K.J. Gregory and D.E. Walling (1968), V.T. Chow (1969), G.H. Comer and R.C. Zimmerman (1969), G.H. Dury (1969), A. Feniran (1969), P. Hagget and R.J. Chorley (1969), J.F. Daniel (1970), M.G. Wolman (1967), V. Gardiner (1971, 1973, 1974, 1976, 1977, 1978, 1979, 1981, 1982), W.B. Kirchner (1975), D.M. Mark (1974), P.M. Mather (1976), P. Mather and J.C. Doornkamp (1970), R.M. McCoy (1970), P.C. Patton and V.R. Baker (1976), K.S. Richards (1979), D.E. Walling (1971), J.S. Smart (1972), J.S. Smart and A.J. Surkan (1967), K.J. Gregory and D. E. Walling (1968, 1971, 1973), K.J. Gregory (1968, 1976) etc.

Drainage basin is a fundamental unit of landform in a fluvial terrain having a particular relevance to the morphometric analysis. Morphometric analysis applied to numerical examples of landforms which may be more properly termed as geomorphometry. The Kosi river basin has been divided into 7th order of intrabasins which form their own ecosystem to regular their self sustained cycle of ecology (Lal, 1988).

The nature and intensity of drainage activities has helped to understand the morphology of the curved landscape.

Morphology of a landscape is a function of drainage, climate and structure of a basin, rock is the base, drainage is the first agent and climate is the catalyst. All these make suitable condition to create appropriate morphology. Thus, 'The analysis of drainage is a very essential to work out its pattern, evolution and genesis and geomorphometry which provide rational development of landscape (Lal, 1998). Drainage basin provides a ideal areal unit within which the collection, processing, organisation and interpretation of data of the geometry of landform, particular in erosional origin can be accomplished. It has remained the geomorphologists objective right from Fenneman (1914 to the present time 2013). 'The limited, convenient and usually clearly define an unambiguous topographic unit available in a nested hierarchy of sizes on the basis of stream ordering and an open physical system in terms of puts of precipitation and solar radiation (Lee, 1964) of the drainage basin confirm' R.J. Chorley (1969) to accept it as a fundamental geomorphic unit from the geomorphometric analysis. Several eminent geomorphologists have accepted a drainage basin for the geomorphmetric analysis.

Geomorphic interpretation and analysis requires quantification of the characteristics and dimensional feature of a landscape.

The Kosi river basin hereby, has been taken as a geomorphic unit for analyzing geometrical properties of the fluvially eroded unit for analyzing quantitatively. 'The need for precise description of geometry of the landforms, particularly those of dominantly fluvial erosive origin, has been recurring theme in geomorphology and one of the most important aspects of this has been the search for the basic areal units within which data could be collected, organized and analysed (Clark, 1966).' The basin elements of morphometric analysis of a basin in a fluvial erosion system are:-

i) The linear properties of the stream channel system
ii) The areal properties of a drainage system
iii) The relief properties of the fluvial system
iv) The slope or gradients of the ground surfaces or stream channels.

For a systematic classification and analysis of geomorphometry of the drainage basin and channel system of the Kosi river basin all the above aspects have been sequentially analysed.

LINEAR ASPECTS OF SAPT KOSI RIVER BASIN : The linear aspects deals with the hierarchical order of streams numbers and length of stream segments and various relatively among them and related morphometric laws e.g. laws of stream numbers and stream length. So, this is undimensional properties of a stream channel system and the branching system of lines are analysed.

The linear aspects of a steam networks are analysed viewing two main points.

i) The topological, which considers the interconnections of the system and yields some scheme of stream ordering.

ii) Geometrical properties, deals with the length shapes and orientation of the constituent parts of the network (Clark, 1973, p.38).

Table - 5.1
LINEAR GEOMORPHOMETRIC PROPERTIES*

S.N.	*LGP	Symbol	Dimension	Related Characteristics Studies
01.	Stream order	(μ)	L	Stream hierarchy
02.	Stream number	($N\mu$)	L	Stream hierarchy
03.	Bifurcation ratio	(R_b)	L	Channel properties
04.	Stream length	($L\mu$)	L	Channel network
05.	Steam width	(W)	L	Rectangular equivalent
06.	Length Ratio	(Rl)	L	Aspect ratio
07.	Length of	(Lg)	L	Aspect ratio
08.	Sinuosity		L	Basin shape
09.	Basin parameter	(P_B)	L	Basin shape

Source: Garde, R.J., 2006: River Morphology, p. 396.

STREAM ORDERING: Stream order refers the determination of the hierarchical position of a stream within a drainage basin. The stream order of Sapt Kosi river has been determined following the Horton's (1949) and modified by the Strahler (1957) and some other scheme of ordering such as Gravielius's scheme, Shreve's stream-link magnitude method, Scheidegger's scheme of stream ordering. 'Stream order is defined as a measurement of the portion of a stream in the hierarchy of a tributary (Leopold, Woldman & Miller, 1969).' Here it is seen that Sapt Kosi river is a 7th order stream according to Strahler's method. Here, it may be mentioned that the hierarchical order increases only when two stream segments of equal order meet and form a junction. The order doesn't increase if a lower stream segment meets a stream segment of higher order. Strahler's scheme is

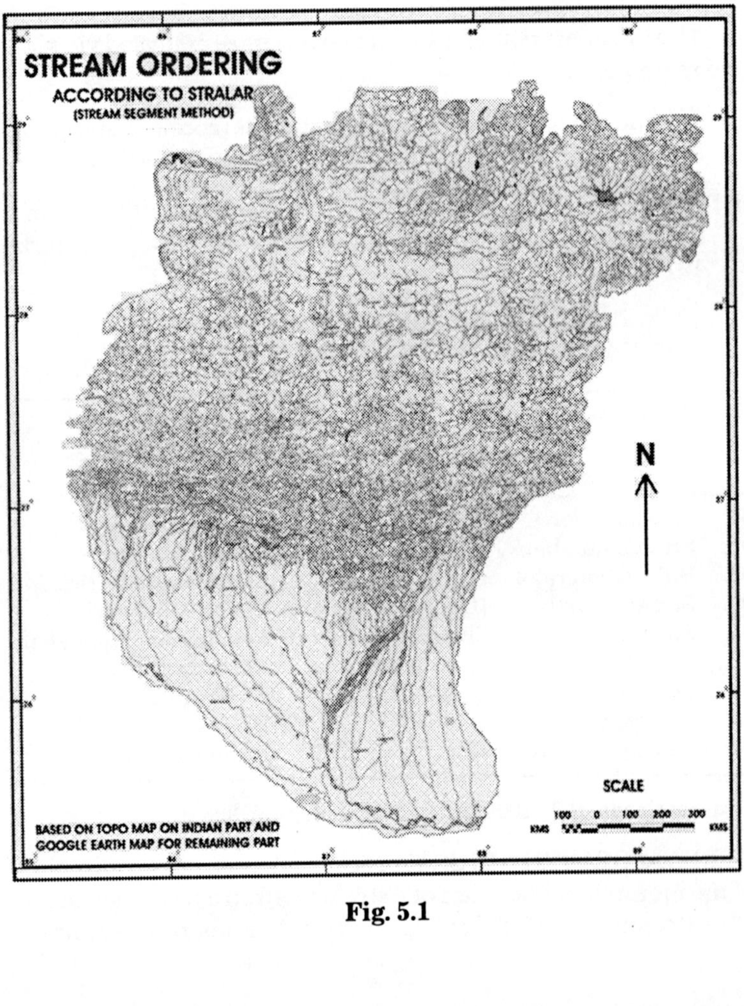

Fig. 5.1

popularly known as stream segment method. A combination of two streams of lower order, say (Nμ-1) with a stream of given order (μ) increases the order of the letter by 1 integer that is μ+ 1. The result of this stream ordering is that it doesn't reflect any increment except approximate by doubling the discharge if one assumes that the stream of the same order of the same drainage basin carrying approximately equal discharge (Fig. No. 5.1).

Some dimensionless features and the weighted mean of bifurcation ratio were calculated. It provides a sequence of stream order numbers which approximate an inverse geometric series for a basin with the degree of branching and bifurcation ratio.

As per the Strahler's methods of stream ordering, the stream order of Sapt Kosi river and its small tributaries have been determined (Table no. 5.2). The Kosi river basin forms 7th order while Bhote Kosi 6th order in which numbers of 1st order stream constitutes 379, 2nd order 82, 3rd order 16, 4th order 4, and one 5th order stream; the numbers of 1st order streams varies from the main channel of Sapt Kosi river basin to the other tributaries of the river basin. The number of highest 1st order stream have been determine for the Kosi river basin while it gradually decreases in the other first order streams likewise in all the tributaries of the basin the number of next higher order also decreases as shown in the above said table.

Table - 5.2
Stream Ordering

Stream order of Sapt Kosi river basin		
Stream order (μ)	**Number of stream segments** (Nμ)	**Bifurcation ratio** (R_b) =
1st order	4,649	4.2
2nd order	1,095	4.6
3rd order	235	5.1
4th order	46	3.1
5th order	15	3.7
6th order	04	04
7th order	01	-
	Average bifurcation ratio	**4.1**

Stream order of Arun river

1st order	1,960	4.1
2nd order	474	4.9
3rd order	97	5.4
4th order	18	3.6
5th order	05	2.5
6th order	02	02
7th order	01	-
	Average bifurcation ratio	**3.8**

Stream order of Tamur river

1st order	531	4.2
2nd order	126	4.3
3rd order	29	4.8
4th order	06	03
5th order	02	02
6th order	01	-
	Average bifurcation ratio	**3.7**

Stream order of Bhote Kosi river

1st order	379	4.6
2nd order	82	5.1
3rd order	16	04
4th order	04	04
5th order	01	-
	Average bifurcation ratio	**3.5**

Stream order of Tama kosi

1st order	96	1.9
2nd order	51	4.6
3rd order	11	5.5
4th order	02	02
5th order	01	-
	Average bifurcation ratio	**3.5**

Stream order of Dudhi Kosi

1st order	206	2.54
2nd order	81	5.8
3rd order	14	07
4th order	02	02
5th order	01	-
	Average bifurcation ratio	**4.3**

Stream order of Bagmati river

1st order			304
3.3			
2nd order	91	5.0	

3rd order	18	4.5
4th order	04	04
5th order	01	-
	Average bifurcation ratio	**4.2**
	Stream order of Kamla Balan river	
1st order	265	5.1
2nd order	52	4.7
3rd order	11	5.5
4th order	04	2.0
5th order	01	-
	Average bifurcation ratio	**4.3**
	Stream order of Sapt Kosi river	
1st order	96	5.1
2nd order	70	4.7
3rd order	15	5.5
4th order	03	2.0
5th order	02	-
6th order	X	-
7th order	01	X
	Average bifurcation ratio	**4.3**

Source : Calculated and analyse by Researcher.

The morphometric analysis of drainage basins carried out by Horton's (1932 & 1945), Strahler (1952), Rzindsym (1960) and other is based on the premise that for the given condition of the lithology, climate, rainfall and other relevant parameter in the basin. The river net, the slope and surface relief tend to reach a steady straight in which morphology is adjusted to transmit the sediment and access flow produced in case of measure climatic and hydrologic discharge in the region the steady straight of the morphologic characteristics naturally modified. In other words 'river net is the definite response of the drainage basin to the complex physical processes taking place over the drainage basin (Garde, 2006).'

BIFURCATION RATIO: Bifurcation ratio is the ratio between the numbers of the stream in a given order and the number of stream in the next higher order. It is a quantitative relation between streams of different order.

Bifurcation ratio (R_b) related to the branching pattern of drainage network and referred as ratio of the numbers of streams of a given order (Nu) to the number of next higher order (Nu+1) and is expresses in the term of the following equation.

$$Rb=\frac{N\mu}{N\mu+1}$$

The bifurcation ratio of the main drainage of Sapt Kosi river basin and its tributaries has been calculated and shown in Table no. 5.2. The bifurcation ratio tends to decrease with increasing order because as order increases the percentage of streams that coalesce into a higher order tributary also increases. This is because of diminishing amount of area available. It is in all the orders the nature of streams register trends of consistence decrease with increase of order but on an average decreasing tend of bifurcation ratio with increasing order. The mean bifurcation ratio of Sapt Kosi river basin is varying from about 4.07 for the plain area in India and 3.12 to 5.1 for mountain, hilly the dissected valley. It is also evident that the mean bifurcation ratio in the range above 5.0 determines to be the basins developed over hilly and highly dissected region. It may be postulated that the mean bifurcation ratio show high variation from mountainous region to the plain region. This variation is described as 'Chance Variations'. Thus the bifurcation ratio depicts that the basin region shows

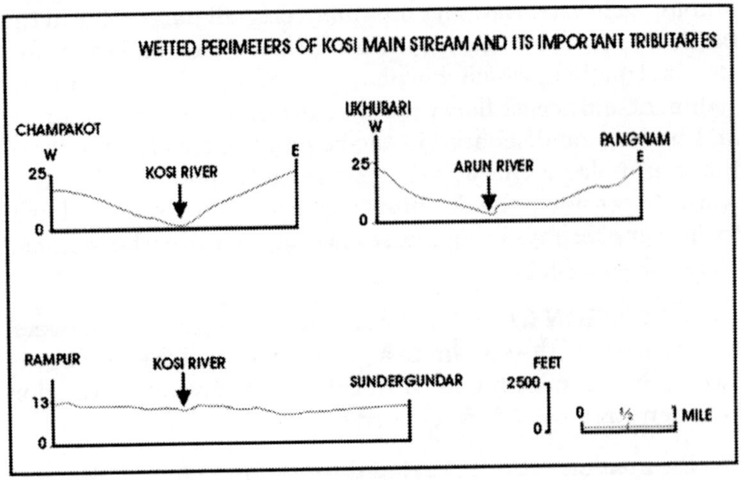

Fig. 5.2

pronounced variation. Owing to Chance Irregularities bifurcation ratio between successive pair of orders, differs within the same basin over it a general observation of a geometric series exists. Therefore, to arrive at a more representative bifurcation number Strahler (1953) used a wetted mean bifurcation ratio. 'The wetted mean bifurcation ratio is obtain by the multiplying the bifurcation ratio for each successive pair of the order by the total number of streams involved in the ratio and taking the mean of the same of the values (Schumm, 1956).' The wetted mean bifurcation ratio of each of the tributaries of the Kosi river basin has been shown in the table no. 5.3.

The wetted mean bifurcation ratio shows great variations the values ranges between 2.8 to 10.8 are indicative of variation in rock type and structures. These values also reflects the ground water situation particularly law order stream in the upper reaches of the basin high value of wetted mean bifurcation ratio indicates shallow ground water, medium infiltration capacity and high surface runoff, particularly in the upper reaches of Sapt Kosi river basin.

Table - 5.3
WETTED MEAN BIFURCATION RATIO OF SAPT KOSI RIVER

Stream order	Number of stream segments	Bifurcation ratio	Number of stream involved in ratio	Product of column 3 & 4
(μ)	($N\mu$)	(R_b)		
1st order	4,649	4.2	5,744	24,124.8
2nd order	1,095	4.6	1,330	6,118
3rd order	235	5.1	281	1,433.1
4th order	46	3.1	61	189.1
5th order	15	3.7	19	70.3
6th order	04	04	5	20
7th order	01	-	-	-

Σ of Column no. 4 = 7,440; Σ of Column no. 5 = 31,955.3; Wetted mean R_b =4.29

Wetted mean bifurcation ratio of Arun river

1st order	1,960	4.1	2,434	9,979.4

2nd order	474	4.9	571	2,797.9
3rd order	97	5.4	115	621
4th order	18	3.6	23	82.8
5th order	05	2.5	7	17.5
6th order	02	02	3	6
7th order	01	-	-	-

Σ of Column no. 4= 3,153; Σ of Column no. 5= 13,504.6; Wetted mean R_b= 4.3

Wetted mean bifurcation ratio of Tamur river

1st order	531	4.2	657	2,759.4
2nd order	126	4.3	155	666.5
3rd order	29	4.8	35	168
4th order	06	03	8	24
5th order	02	02	3	6
6th order	01	-	-	-

Σ of Column no. 4 = 858; Σ of Column no. 5 = 3,623.9; Wetted mean R_b = 4.2

Wetted mean bifurcation ratio of Bhote Kosi river

1st order	379	4.6	461	2,120.6
2nd order	82	5.1	98	499.8
3rd order	16	04	20	80
4th order	04	04	5	20
5th order	01	-	-	-

Σ of Column no. 4 = 584; Σ of Column no. 5 = 2,720.4; Wetted mean R_b = 4.6

Wetted mean bifurcation ratio of Sun Kosi river

1st order	812	11.9	880	10,472
2nd order	68	2.8	92	257.6
3rd order	24	4.8	29	139.2
4th order	05	2.5	7	17.5
5th order	02	02	3	6
6th order	01	-	-	-

Σ of Column no. 4 = 1,011; Σ of Column no. 5 = 10,892.3; Wetted mean R_b = 10.8

Wetted mean bifurcation ratio of Tama Kosi river

1st order	96	1.9	147	279.3
2nd order	51	4.6	62	285.2
3rd order	11	5.5	13	71.5
4th order	02	02	3	6
5th order	01	-	-	-

Σ of Column no. 4 = 225; Σ of Column no. 5 = 642; Wetted mean R_b

= 2.8

Wetted mean bifurcation ratio of Dudhi Kosi river				
1st order	206	2.54	287	729
2nd order	81	5.8	95	551
3rd order	14	07	16	112
4th order	02	02	3	6
5th order	01	-	-	-

Σ of Column no. 4 = 401; Σ of Column no. 5 = 1,398; Wetted mean R_b = 3.5

Wetted mean bifurcation ratio of Bagmati river				
1st order	304	3.3	395	1,303.5
2nd order	91	5.0	109	545
3rd order	18	4.5	22	99
4th order	04	04	5	20
5th order	01	-	-	-

Σ of Column no. 4 = 531; Σ of Column no. 5 = 1,967.5; Wetted mean R_b = 3.7

Wetted mean bifurcation ratio of Kamla Balan river				
1st order	265	5.1	317	1,616.7
2nd order	52	4.7	63	296.1
3rd order	11	5.5	13	71.5
4th order	02	2.0	3	6
5th order	01	-	-	-

Σ of Column no. 4 = 396; Σ of Column no. 5 = 1990.3; Wetted mean R_b = 5.0

Wetted mean bifurcation ratio of Sapt Kosi river				
1st order	96	5.1	166	846.6
2nd order	70	4.7	85	399.5
3rd order	15	5.5	18	99
4th order	03	2.0	4	8
5th order	01	-	-	-
6th order	x	-	x	x
7th order	01	-	x	x

Σ of Column no. 4 = 273; Σ of Column no. 5 = 1353.1; Wetted mean R_b = 4.96

Source : Calculated and Analyse by Researcher.

Table - 5.4
LENGTH RATIO AND LAW OF STREAM LENGTH

Stream Order (μ)	Stream Segments ($N\mu$)	Constant Bifurcation Ratio (R_b)	$\sum N\mu = R_b \frac{k-1}{Rb-1}$
1st order	4,610	4.7	31.7
2nd order	970	4.5	30.3
3rd order	214	3.8	25.3
4th order	56	4.3	28.8
5th order	13	3.25	21.4
6th order	04	04	26.7
7th order	01	-	-
Average bifurcation ratio		**4.1**	**27.4**

Source: Computed from Topomap no. 72/G & 72K (for Indian part) and Google earth images (for China & Nepal part).

LAW OF STREAM NUMBERS: The law of stream numbers related to the definite relationship between the order of the basin and the stream numbers. 'Stream number states that the numbers of stream segments of successfully lower order in a given basin tend to form a geometric series beginning with the single segment of the highest order and increasing according to constant bifurcation ratio (Hortons, 1945).' The law of stream number is expressed by the negative exponential function model $Nu=R_b^{(k-\mu)}$ considering the net of a drainage basin in which the highest order of the stream is K. Let, μ represent of the any segment and $N\mu$ represent of the streams of the order μ. as per Horton if Log ($N\mu$) is plotted against μ for any river. Net the data fall on a single straight line with $N\mu$ decreasing for increasing μ. Here the relation between $N\mu$ and μ can be expressed as the below. The stream number of the Sapt Kosi river basin has been computed (Table no. 5.4).

The principal embodied in equation is commonly known as Horton's law of stream number which states that 'the number of streams of different order of a given drainage basin trends closely to approximate and inverse geometric series in which the first term is unity and ratio is the bifurcation ratio.'

LENGTH RATIO AND LAW OF STREAM LENGTH: 'the average length of the each of the different orders in a drainage basin tend closely to approximate a direct geometric series in which the first term is the average length of stream of 1st order. The order wise length has been finding with a rotameter. The total length of the all stream of given order is the predominant of the number of streams and the average length. The mean length $L\mu$; cumulative mean lengths ($CL\mu$) and length ratio (R_L) were calculated the date plotted on algorithm graphs. The stream length ratio has been completed in Table no. 5.5

Table - 5.5

STREAM LENGTH AND LENGTH RATIO

Stream Order	Stream Number	Stream Length	Mean Stream Length	Cumulative Stream Length	Length Ratio
(μ)	($N\mu$)	(In km)	($L\mu$)	($CL\mu$) in km	(R_L)
1st Order	4,610	17,641	4	4	-
2nd Order	970	7,147	7	11	3
3rd Order	214	3,493	16	22	9
4th Order	56	1,892	34	50	18
5th Order	13	1,103	85	119	51
6th Order	04	871	218	303	133
7th Order	01	857	857	1,075	639
Total/Mean	**5868**	**33,004**	**174.43**	**226.28**	**142.17**

Source: Computed from Topomap no. 72/G & 72K (for Indian part) and Google earth images (for China & Nepal part)

The average stream length ratio (Table no. 5.6) of different streams of Sapt Kosi river basin is:

Table - 5.6

STREAM LENGTH RATIO OF DIFFERENT DRAINAGE BASINS

S.N.	Name of the River Basin	Stream Number ($N\mu$)	Stream Length in km	Mean Stream Length ($L\mu$)	Average Length Ratio (R_L)
01	Arun	2,557	28,256	11.05	-
02	Bhote Kosi	482	2,199	4.56	- 6.49
03	Tama Kosi	161	1,323	8.22	3.66
04	Dudhi Kosi	304	1,517	4.99	- 3.23
05	Sun Kosi	912	3,825	4.19	- 0.8
06	Tamur	695	2,810	4.04	- 0.15
07	Bagmati	418	2,020	4.83	0.79
08	Kamala Balan	333	2,198	6.60	1.77
09	Sapt Kosi	187	2,999	16.04	9.44
Total/Mean		**6,049**	**47,147**	**7.79**	**0.62**

Source: Computed from Topomap no. 72/G & 72K (for Indian part) and Google earth images (for China & Nepal part)

It is evident from the table that the stream length ratio of the different drainage basins of Sapt Kosi river basin ranges from -6.49 to 9.44. The mean stream length plots adhere to the Horton's law of stream length. The plots of logarithm of stream length as a function of stream added yields a set of points line essentially along the straight line as it is appropints from the absolute length of the channel system exercises the control of the basin. it is hereby pertain to maintain that where as 'Horton's (1945) found that the value of R_b varied from 2 to 4 for the river nets investigated by him where as Strahler's scale ranges to be from 3 to 5 for the drainage basins in which geologic structure do not destroyed the drainage pattern.' Mittal, at. el., 1974 found R_L between 4.4 and 4.0 for most of the drainage basins of the Garhwal Himalaya, were lithologic conditions were represented by limestone, quartzite, Philites, shale, sandstone and alluvium. In case of Sapt Kosi river basin the length ratio has been observe up to 0.62. In view of orderwise the length ratio is increasing gradually. In its upper reaches formed by sandstone, limestone and quartzite, it is indicative of the weak lithologic conditions. The absolute length of the channel system exercises minimum control over the basin.

SINUOSITY INDICES: 'No straight river occurs in a straight of nature (Leopold & Wolman, 1960).' Some of the rivers are meandering and all are sinuous. 'Meandering is applied to sinuous channels that exhibit a certain degree and regularity of sinuosity (Brice, 1962).' A drainage line involve in the calculation of deviation of observe path (OL) from the expected path almost a straight line (EL) of a river from the source to the mouth. Sinuosity of a stream denotes a degree of deviation of its actual path from the expected theoretical path. The degree of sinuosity reflects the stage of basin development and landform development as well.

The valley of unity (1.0) of standard sinuosity index HSI is indication of straight river course, where as valleys between 1.0 to 1.5 reflects sinuous shape and more than 1.5 represents a meandering course. The HSI and TSI has been determine from the relevant topographic sheets in which HSI indicates the proceeding of streams departure from a straight line course is due to hydraulic differences from the straight line course. On the above all the streams of Sapt Kosi river basin measured are in the juvenile stage. The HSI and TSI have been calculated in Table no. 5.7.

Table - 5.7
SINUOSITY INDEX

Stream Order	CL	VL	AL	CI	VI	HSI	TSI	SSI
(μ)	In kilometers			$\frac{CL}{AL}$	$\frac{VL}{AL}$	$\frac{CI-VI}{CI-1}$ x100%	$\frac{VI-1}{CI-1}$ x100%	$\frac{CI}{VI}$
1st order	17,641	16,441	8,175	2.16	2.01	14.8	87.1	1.07
2nd order	7,147	6,841	4,323	1.65	1.58	10.8	89.2	1.04
3rd order	3,493	3,224	1,645	2.12	1.96	14.3	85.7	1.08
4th order	1,892	1,674	984	1.92	1.70	23.9	76.1	1.13
5th order	1,103	965	583	1.89	1.66	25.8	74.2	1.15
6th order	871	847	335	2.6	2.53	4.4	95.6	1.14
7th order	857	841	372	2.30	2.26	3.1	96.9	1.02
					Mean	**13.87**	**86.4**	**1.09**

Source: Computed from Topomap no. 72/G & 72K (for Indian part) and Google earth images (for China & Nepal part)

The values of HSI and TSI obtained for Sapt Kosi river basin ranges between 3.1 to 25.8 and 74.2 to 96.9 % which is more than 1.0 and represents meandering course. It is also indicative of stage of basin development. Most of the streams having HSI value between 3.1 to 25.8 which shows comparatively juvenile stage of basins development. TSI dominants over HSI and after the removal of pronounced relief control in the late mature and old stage. Hydrological sinuosity (more than 60%) scores over topographical sinuosity. Thus, the sinuosity indices are enough to explain the hydrological and topographical characteristics of Sapt Kosi river basin.

MEANDER PROPERTIES: Meander is an important geometric properties which explains the nature of the course. Many fluid threads other than river are known to meander. Thus, meandering is therefore the normal aspect of a fluvial thread within certain velocity and viscosity limits, unless hindered by some outside influence. Meander traditionally ascribed as fluvial plains and delta plain.

The high degree of symmetry of meander length, meander height and form ratio is the characteristics of meandering properties which are applied to a sinuous stream. That has, from place to place along its course, one or series of symmetrical arcs, the length of which is related with the width of the stream (J.C. Brice, 1964). Brice has developed a model for computing meander properties. It involves calculation of symmetry of meander length, height, form ratios and repetition of symmetry, which as follows.

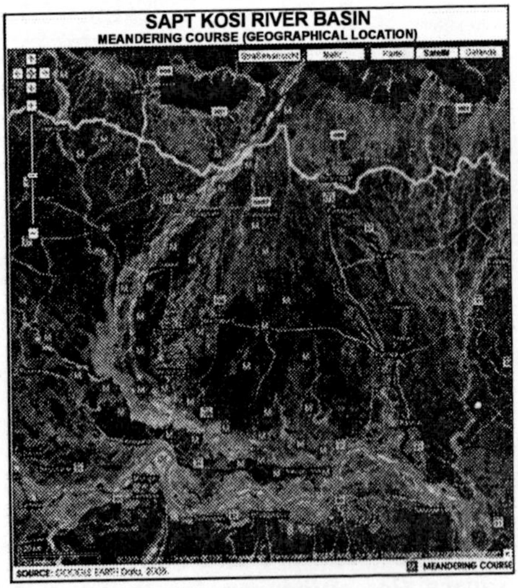

Fig. 5.3

i) Symmetry of meander length
ii) Symmetry of

meander height

iii) Symmetry of form ratio,

The symmetry of Sapt Kosi river basin has been calculated more than 80% in the plains of Bihar and above 60% in mountainous area which indicates towards the meandering course of the Kosi river. Sapt Kosi river basin in the lower reaches in the plains of Bihar is highly meandering. (Fig. No. 5.3) While the repetition of symmetry is a vital significance in deciding the meandering properties of the streams, where the higher percentage of repetition of symmetry is indicates towards meandering course. The higher symmetry has been found in the plain tract of Bihar.

Meanders are the result of the erosion and deposition both.

i) On the basis of nature of fluvial erosion there are two major type:

a. Meander develop through lateral erosion

b. Meander develops through vertical erosion (incised meander)

Lateral erosion is more common in the plain tract of the lower Sapt Kosi river basin. It is erosional meander, where as the incised meanders are generally found in the middle and upper reaches of Sapt Kosi river basin. Although, morphologically meanders are of three types:

i) Weavy type of meander
ii) Horse shoe type of meander
iii) Ox-bow bracelet type of meander

Weavy meanders are simple in plain and it is mostly develop in the various stream of Sapt Kosi river. It is also pertinent to say that simple meanders develop during the juvenile stage of fluvial cycle of erosion. The upper and middle reaches of Sapt Kosi river basin shows juvenile stage and the meanders are found mostly weavy and horse shoe type; whereas the incised meanders developed due to vertical erosion and forms five types such as incise meander, enterace meander, interched meander, enclosed meander and ingrown meander. In the upper reaches of Sapt Kosi river basin meander are mostly inclosed and incised

Table - 5.8

GEOMETRY OF MEANDERS OF DIFFERENT CHANNEL SEGMENTS

Channel and Location	Meander Length	Mean Width	Arc Distance	Mean Radius	Channel Width	Amplitude (According to Leopold & Wolman)	Sinuosity
	(meter)	(meter)	(meter)	(rm)	(meter)	$A = 2.7w^{1.1}$	SSI =
Sapt Kosi, Kursela	6000	3000	3000	4000	600	3392	0.75
Sapt Kosi, Naugachhia	5000	3000	3000	3000	600	3392	0.75
Kamla Balan, Chautham, Khagaria	2000	950	1500	1240	570	2903	1.13
Tilabeh, Saharsa	3800	2200	1900	2100	500	2513	1.15
Bagmati, Bachauli, Darbhanga	4100	2050	2900	1300	580	2959	1.14
Arun, Triveni	373	388	186	170	310	1485	2.12

Source: Computed from Topomap no. 72/G & 72K (for Indian part) and Google earth images (for China & Nepal part)

characterised by deep and narrow valleys. Entraced and interaced meanders are found to occur in the lower reaches of the basin and result of lateral erosion.

The quantitative measurement of the meander curves has been done by many geomorphologists viz. Inglis, 1937, 1949; Jafferson, 1902; bates, 1939; Leopold & Wolman, 1957, 1960. The relationship between meander length, amplitude, radius of curvature and channel width are consider to be linear (Leopold & Wolman, 1957).

It is obvious from the table (5.8) that the values of the meander (sinuosity) in all the streams of Sapt Kosi river basin are under 0.75 to 2.12. The meander in the stream of the hilly region is in between 1.14 to 2.12. It is sinuous stream. The development of pools and riffles produces undulating channel bed which causes oscillation in flow velocity downstream. Lateral oscillation of channels increases sinuosity index and development of moderate bend in the channel. It may be mention that pools favor the formation of bend which straight reaches develops the riffles. As per K. S. Richard (1982) development of meander are the mostly the result of third stage based on hydraulic principles. The third stage is heralded by the development of the ideal meander belt, where in one meanders wavelength is equal to 10-12 mean meander width. It is also important to mention here that 'ones a meander pattern is established it is likely to précised unless some really powerful disturbing factor comes into operation (Singh, S., 2011; Reprinted).' But although a great deal is known about the behavior of meandering stream and about their statistical properties. It is still not possible to define the ultimate cause of meandering (Dury, 1969).' Meander wavelength has been found to be depended on average channel width, radius of metamorphic curvature and dominant discharge. It has been observed that the meander wave length is moderately correlated with discharge but very strongly correlated with channel width. (L= 10-14w; W= mean channel width) S.A. Shumm (1968) has established that meander wavelength (L) is 'the formation of both bankful discharge (Qb) and the wetted percentage of silt-clay in the benchmark material' (m) (Petes & Foster, 1985) $L= 618Qb^{0.43}M^{-0.74}$ on the above basis the greater wavelength of sand and gravel transporting rivers than silt-clay transporting rivers.

Sapt Kosi river basin is the lower catchment area forms wider beds of meander. It is the result of the dominance of hydrodynamic factor and irregularity is cause by random topographical, sediemntological or artificial disturbances flow of water in the meander channel of Sapt Kosi particularly in the lower catchment area takes on a spiral course it is like a screw thread, and this trends to maintain the meandering form.

Table - 5.9

LENGTH OF OVERLANDFLOW

Drainage Basin	Stream Frequency (Fs) $Fs = \frac{Ns}{A}$	Drainage Density (Dd) $Dd = \sum\frac{L}{A}$	Length of Overland Flow (Lg) $Lg = \frac{1}{2Dd}$
Arun	13.55	1.22	0.61
Bhote Kosi	13.98	3.1	1.55
Tama Kosi	37.12	4.52	2.26
Dudhi Kosi	9.56	1.91	0.95
Sun Kosi	4.23	1.01	0.50
Tamur	5.76	1.42	0.71
Bagmati	32.51	6.73	3.36
Kamala Balan	17.02	2.58	1.29
Sapt Kosi	75.67	4.69	2.34

Source: Computed from Topomap no. 72/G & 72K (for Indian part) and Google earth images (for China & Nepal part).

LENGTH OF OVERLAND FLOW: Overland flow is consider as a dominant hydrological and morphologic factor is the mean horizontal length of flow path from the divide to the stream in a first order stream and is a measure of stream spacing and degree of dissection and is approximately one half the reciprocal of the drainage density (Chorley, 1961). It is an important morphological variable which affects the hydrological and topographical development of the basin. It also related to the stage of development but no definite parameter or limit or length of overland flow has been suggested. It is generally, a quantitatively closer that the early stage is marked with maximum length of overland flow and mature and old stage register mark reduction in Lg (Table no. 4.9).

'Overland flow may be wholly turbulent, wholly laminar or partly turbulent and partly laminar (Horton, 1945).' 'The relation between depth of surface detention and runoff intensity, In inches per hour should be a simple power function of the depth of surface detaintions (Schumm, 1956).' The length of overland flow is higher in Sapt Kosi river basin due to steep slope and greater undulations of Sapt Kosi catchment area. Length of overland flow is very small in the mountainous tract of the Himalayan region, the nature of the valley characterised by deep gorges and small width of channel whereas length of overland flow is higher in the lower catchment area of Sapt Kosi river basin. The length of overland flow of entire Sapt Kosi river basin is quit varies. It is 0.61 in the stream of the upper catchment area and 3.36 in the plain tract of the basin. The overland or sheet flow is very remarkable in the plain tract. It is commonly seen that the upper mountainous track is highly venerable to the soil erosion in comparison to the lower catchment area. Sheet flow to some extent but deposition of sediment, like silt, clays and to some extent gravels is high in the lower catchment area.

STREAM JUNCTION ANGLE: Angle property of a drainage stream is related to the junction angle of two stream segments and referred as the angle, projected to the horizontal, between average flow direction, determine by the ends of the stream segment, extended from the junction point to an upstream point at a distance equal to 0.2 times and average length of the second order stream (Lubowe, 1964). 'A traditionally to channel of higher order develops with an entrance angle dependent on the ratio between channel and ground slope (Schumm, 1956). 'Where the valley side slope is steep in relation to the gradient to the master stream and valley side slope are the almost same, the tributary vertically parallel the main channel, joining it at small angle.

'Because the junction angle is the function of the ratio of the stream gradient, a higher angle can be expected when a low order stream joints one of the higher angle and the angle trading to approach 90^0 as the difference between order increases (Lubowe, 1964).' It is evident that the mean junction angle increases up to 90^0. As the order of the receiving streams increases. Limiting condition which is approach but not often attain (Horton, 1945). the stream junction angle of most of the streams of the Kosi river basin reflects the basin development. Most of the tributary streams of Sapt Kosi river basin are of 4th order and 6th order in this three river mention in the junction angles have been measured up to 6th order. It is observed from the basis that in these rivers.' The junction of the 1st order streams with the 4th order and the junctions of the 2nd and 3rd order streams with the stream of the same order or the higher order are uniformly over 80^0 and in all of them the mean junction between 3rd and 4th order stream is approaching 90^0.

RELATION BETWEEN STREAM ANGLES TO GRADIENT: A clear relationship exists between the stream junction angle and the angle predicted by Horton's equation which is predicted and observe mean angle area shown in the table no. 5.10.

Table - 5.10

PREDICTED AND OBSERVED JUNCTION ANGLE

Drainage Basins	Gradient Ratio	Horton's Predicted Junction Angle	Observe Junction Angle	
Rivers	00:00	$\text{Cos zc} = \dfrac{\tan sc}{\tan sg}$	in ⁰ as horizontal angle	
Arun			Arun	
	4:3	1.34	Far Upper	S 15^0W
	9:7	1.29	Upper	
	8:6	1.34	Middle	S 45^0W
	8:6.3	1.27	Lower	S 20^0E
	5:4.3	1.16	Lowest	S 40^0E
Bhote Kosi	9:5	1.8	S 30^0E	
Tama Kosi	9:6	1.5	S 30^0E	
Dudhi Kosi			Dudhi Kosi	
	7:5	1.4	Upper	N-S
	6:5.3	1.3	Lower	S 65^0W
Sun Kosi	6:1.1	1.17	Upper	S 30^0W
	8:7	1.14	Middle	S 30^0E
	7.6.1	1.14	Lower	S 40^0E

	6:5.4	1.11	Lowest	S 70⁰E
Tamur	4:3	1.34	Upper	S 45⁰W
	4:3.3	1.21	Middle	S 45⁰E
	3:2.6	1.15	Lower	S 45⁰W
	3:2.7	1.11	At Dhanukta	
	3:2.9	1.03	Lowest	S 75⁰W
Bagmati	2:1.9	1.05	S 12⁰W	
Kamala Balan	2:1.7	1.18	N-S	
Sapt Kosi	8:7.3	1.09	Kosi in george	S 50W
	7:6.3	1.11	Kosi in Terai	
	6:5.6	1.07	Kosi in Fan	S 45⁰W

Source: Computed from Topomap no. 72/G & 72K (for Indian part) and Google earth images (for China & Nepal part).

The junction angle is more or less similar to Horton's predicted junction angle. The average gradient of the first order stream the ratio of tan Sc and tan Sg or the gradient ratio is observe of the stream is similar to the Horton's prediction. There doesn't see to take any direct relation between lithology and stream junction angle. It is relevant to mention here that lithology doesn't control any controlling influence over the junction angle. Similar axile angle have been observed in the various lithology where as junction angle differ considerably in homogonous rocks. It is obvious from the table (5.10) the maximum angle of direction of the stream has been observing the Tamur at Dhankutta lowest. It is 75⁰W. Some other streams are also having more than 50% of stream junction angle. It shows the stage of the basin development higher the angle of direction is the result of juvenile stage of development.

AREAL ASPECTS: The areal aspects of the drainage basin indicate the study of the basin parameter geometry of close links that is basin shape, law of basin, law of altimetric growth, stream frequency, drainage density, drainage texture etc.

Basin area is very important morphometric attribute as it is directly related to the spatial distribution of a number of significant attributes. H.W. Anderson has remarked it as a Devil's own variable because every watershed characteristics is correlated with area. The computations of the basin area of different order have been devised by the geomorphologists. The drainage basin area is delineated on the basis of water divides and the rocks of the all stream segments of each order are measured with the help of digital planimeter. 'All of the gradient surface, which directly feeds the 1st order segments, are induced the area of the 1st order stream. The area of the 2nd of order stream segments includes the area of the 1st order segments plus the areas of the interbasins which are triangular patches of the ground surface contributing to the directly to the 2nd order. The same principles is adapted to the successive order (Singh & Srivastava, 1974).' It is thus the basin area automatically becomes cumulative from the 1st order to the successive higher order.

The evolution of the hydrographic network is directly related to the distribution of the sources and confluence point which have direct bearing on the geomorphic evolution of the drainage system. The different areal aspects of morphometry of the Kosi river basin are as follows:

BASIN SIZE: The maximum flood discharge per unit area is closely related to the basin size (Moore, 1993). The Sapt Kosi river basin is prone to flood. In calculating the basin are surface are of the watershed, the entire area between the divide line and the outfall with all sub and interbasin area have been considered (Table no. 4.11).

Table - 5.11

BASIN AREA OF THE DIFFERENT STREAMS

S.N.	NAME OF THE STREAM	BASIN AREA IN SQ. KM
01.	Arun	34,660
02.	Tamur	6,741
03.	Bhote Kosi	5,976
04.	Tama Kosi	2,905
05.	Sun Kosi	3,857
06.	Dudhi Kosi	4,005
07.	Bagmati	13,589
08.	Kamla Balan	5,669
09.	Sapt Kosi	14,074
	Total	**91,476**

Source: Computed from Topomap no. 72/G & 72K (for Indian part) and Google earth images (for China & Nepal part)

As Sapt Kosi river basin is compared of complex geologic structure. The sub surface water doesn't play an important role. The Kosi river basin receives water from the rainfall as well as snow melt. The larger the size (area), the greater the precipitation and the higher is the runoff. The Kosi river basin has an extensive basin area with steep slope gradient in the upper and middle reaches which gentle in the plain tract in the lower reaches. The peak discharge is of greater significance in deciding upon the size of the culeverts, cause-ways and bridges.

The orderwise area of the Kosi river basin:

Orderwise area (Table no. 5.12) is an indicative of the amount of discharge of water through the streams.

Table - 5.12
ORDERWISE AREA OF DIFFERENT STREAMS

Stream Order	Total Area	Cumulative Area	Percentage of the Total Area	Mean Area in Sq. Km.	Area Ratio
	In Sq. km	In Sq. km	(%)	($\hat{A}\mu$)	(Ra)
1st Order	27,545	27,545	30.11	5.92	5.92
2nd Order	19,223	46,768	21.01	17.56	11.64
3rd Order	13,856	33,079	15.15	58.96	41.4
4th Order	8,734	22,590	9.55	189.87	130.91
5th Order	7,432	16,166	8.12	495.47	305.6
6th Order	8,554	15,986	9.35	611	115.53
7th Order	6,132	14,686	6.71	6132	5521
Total/Mean	**91,476**		**100%**	**1073**	**876**

Source: Computed from Topomap no. 72/G & 72K (for Indian part) and Google earth images (for China & Nepal part)

It is observed from the above table that percentage of area is much occupied by 1st order of stream (30%) and lowest percentage of area is occupied by 7th order stream (6.7%) which is mainly Sapt Kosi (after Chatra to Kursela) and Arun river.

LAW OF BASIN PARIMETER: Basin perimeter is an important morphometric perimeter which helps in determines the shape, size and genetic aspects of drainage basin. Perimeter of drainage basin is directly related with the roots of basin area and increase or decreases of either two the coefficient correlation of nine tributaries of Sapt Kosi river basin standing at 0.85 which is indicative of strong positive correlation. It is significance of at 1% probability level. Similarly, co-efficient correlation between basin perimeter and channel lengths and square roots of basin area and channel area being 0.78 and 0.85 are again finding out at 1% confidence level. It narrates the genetic aspects of basin development. All of the 1st order of the basin is invulred in the lowering down the watershed of the area through the headward erosion. It has resultant into the continuous growth of the basin perimeter. The analysis of the different watershed area of the basin shows adverse correlation in the upper part of Sapt Kosi river basin where as it is quite positive in the lower catchment area. 'This law is very much apply in almost all the heterogeneous lithology and geologic structure. But significant variations in relief characteristics upsectes smooth functioning of the drainage basin development and departures from the above law (Singh & Upadhaya, 1982).'

AREA RATIO: Area ratio refers proportion of increase of mean basin area between two successive orders.

Table - 5.13
AREA RATIO (Ra)

Stream Order	Number of All Segments of a Given Order ($N\mu$)	Mean Area of Two Successive Order ($\hat{A}\mu$)	Area Ratio $\frac{\Sigma \hat{A}\mu}{N\mu}$
1st Order	4649	5.92	5.92
2nd Order	1095	17.56	11.64
3rd Order	235	58.96	41.4
4th Order	46	189.87	130.91
5th Order	15	495.47	305.6
6th Order	14	611	115.53
7th Order	01	6132	5521

Source: Computed from Topomap no. 72/G & 72K (for Indian part) and Google earth images (for China & Nepal part)

It is observed from the Table no. 5.13 that area ratio of 1st order is 5.92 where as it is 5,521 for 7th order stream. Almost the area ratio is increasing gradually from the lowest to highest level of stream order. It is 131 for 4th order stream and 306 for 5th order stream. It is reflective as the mean basin area between the two successive orders is increasing as the orders increasing and basin area is decreasing. It is proportional to the lower order to the higher orders. Sapt Kosi river basin in the upper most reaches where 1st order streams is mostly exists covers a wide and extensive area where as in the 2nd and 3rd order of stream area where the streams make their shapes area of the basin proportionately squeezes.

LAW OF BASIN AREA: A. N. Strahler (1969) paraphrased Horton's law of stream length into law of basin area and postulated that 'The mean basin areas of successive higher stream orders tend to form geometric series beginning with mean area of the 1st order basin.

In case of the Bhote Kosi river the coefficient correlation is 1.538%. It shows under estimation in basin area of 1st, 2nd and 4th order. It is correlative in the scenes that the ratio within different order of a particular basin becomes more or less stable. The law of basin area near comes to the truth but when area ratio presents high variation the law of basin area losses its validity.

LAW OF ALOMETRIC GROWTH: The successive increasing orders with the advancement of fluvial cycle indicate increasing time where in appearance of new 1st order finger tip stream segments with accelerated rate of erosion during the first phase of youthful stage increases the order of the older segments. In any fluvial system under the idle conditions of uniform structure and climate, the rate of increase becomes orderly and follows of law of alometric growth. Thus, the law of alometric growth states proportionate increase in all part of the organs with living organism with time of organs of animals and plants has been introduced by the geomorphologists M. Morisawa, 1957; M. J. Wolderberg, 1966; A. N. Strahler, 1969.

Since Sapt Kosi river basin characterised by heterogonous geological structure and climatic conditions, more than 60% of the finger tip streams are under the direct glaciated impact. Adverse condition exists as the upper catchment area is under the glacier and periglacial climatic conditions where the middle and lower catchment area is under the fluvial impact. So this law of alometric growth did not fit for the Kosi river basin but to the some extend the tributary streams which get its origin from the middle Himalaya and the lower part of the basin area which comes under the fluvial action and almost the same climatic conditions prevails this part of the basin is under the law of alometric growth. If the cumulative mean length and the mean basin areas of the different river of Sapt Kosi river basin gets rise in the middle and lower reaches are plotted on altimetric graph paper (constant ratio scale). A straight line of regression of positive power function is formed.

On the basis of equation the high degree of correlation of 0.99 and the percentage of variance explain, 99.8% between cumulative mean length and basin area of the Bhote Kosi, Dudhi Kosi, Sun Kosi and Tamur.

STREAM FREQUENCY: Stream frequency or drainage frequency is defined as the numbers of river or more precisely the river segments of all orders in a basin divided by the surface area of the basin. It is measured by the number of streams per unit area.

Table - 5.14
STREAM FREQUENCY

Drainage Basin	Total Number of Streams (Ns)	Total Area in Sq. Km. (A)	Stream Frequency $Fs = \frac{Ns}{A}$
Arun	2,557	34,660	0.07
Bhote Kosi	482	6,741	0.07
Tama Kosi	161	5,976	0.03
Dudhi Kosi	304	2,905	0.10
Sun Kosi	912	3,857	0.24
Tamur	695	4,005	0.17
Bagmati	418	13,589	0.03
Kamala Balan	333	5,669	0.06

Sapt Kosi	187	14,074	0.01
TOTAL	**6,049**	**91,476**	**0.07**

Source: Computed from Topomap no. 72/G & 72K (for Indian part) and Google earth images (for China & Nepal part)

The stream frequency is closely associated with the bifurcation ratio. It is evident from the Table no. (4.14) that the highest stream frequency is of the Sun Kosi (0.24) and it is 0.03 the lowest stream frequency of the Tama Kosi as well as Bagmati which indicate more or less the same lithological character. The stream frequency of Kamla Balan (0.06), Dudh Kosi (0.10) indicates variation in the lithological characteristics.

DRAINAGE DENSITY: 'Drainage density is defined as the total length of streams in a basin divided by its area, hence the drainage density (Dd) is given by Dd = ' (Garde, 2006), Strahler (1964) consider that drainage density (Dd) to be a variable dependent on runoff per unit area Qr, erosion proportionately factor k (which is mass rate of erosion per unit area per unit eroding force), relief H, mass density of fluid $ñ_{5S}$, dynamic viscosity u and g.

Dd = f (Qr, KH, pf, u, g)

On the above equation has been computed on the basis of the regional scale and divided the basins into grid squares and exhibited by Figure no. 5.4.

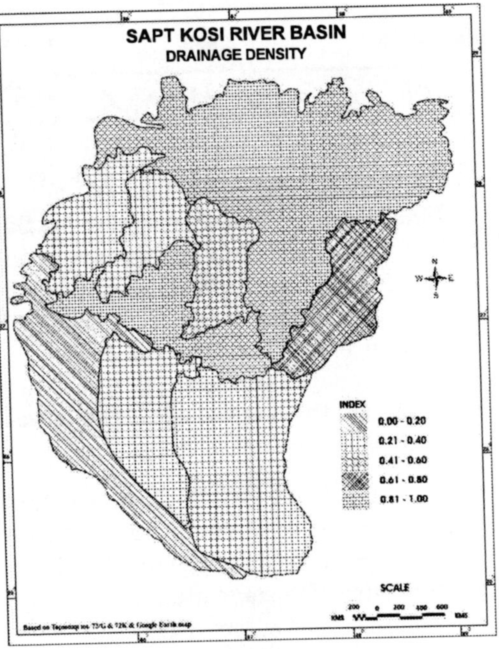

Fig. 5.4

The special variation in drainage density has been

related to the precipitation effectiveness, (M. M. Melton, 1957), vegetation index (R.J. Chorley, 1957), permeability of terrain (C. W. Carlston, 1963), climatic characters (C. A. Cotton, 1964), rainfall intensity (M.A. Morgan, 1962 & M. A. Molton, 1957).

Table - 5.15

DRAINAGE DENSITY

Drainage Basin	Total Stream Length	Total Area in Sq. Km.	Drainage Density
	(L)	(A)	(Dd) $Dd = \sum \frac{L}{A}$
Arun	28,256	34,660	0.82
Bhote Kosi	2,199	6,741	0.33
Tama Kosi	1,323	5,976	0.22
Dudhi Kosi	1,517	2,905	0.52
Sun Kosi	3,825	3,857	0.99
Tamur	2,810	4,005	0.70
Bagmati	2,020	13,589	0.15
Kamala Balan	2,198	5,669	0.39
Sapt Kosi	2,999	14,074	0.21
TOTAL	**47,147**	**91,476**	**0.51**

Source: Computed from Topomap no. 72/G & 72K (for Indian part) and Google earth images (for China & Nepal part).

The average drainage density of Sapt Kosi river basin is 0.51. It varies in different streams of the basin but in all the basins the drainage density is under 1.00 which shows a very meager drainage density. It is due to the high slope of the catchment area. It is also due to the fact that the horizontal erosion is meager and large channels develop to provide quick channels. On the basis of obtain solve value the basin has been divided into four density zone.

DRAINAGE TEXTURE: Drainage texture is meant by 'the relative spacing of drainage lies (Thornbury, 1957).' It includes both the drainage density and stream frequency. It is directly related with drainage density and influenced by climate infiltration capacity, mean available relief, geological structure. No successful attempt has been made to search out a quantitative parameter for the calculation of the drainage texture. The term 'drainage texture must be used to indicate relating spacing of the streams in a unit area along a linear direction (Singh & Srivastava, 1976).'

The drainage texture of Sapt Kosi river basin is very course in the upper reaches which is indicative of the medium drainage texture. The extreme upper most area of Sapt Kosi river basin and its several tributaries registers medium textures where the lower most catchment area the contour spacing are widely spread which shows the low category of texture. The dominance of a complex structure and rocks specially in the Bhote Kosi and Arun river. Lithological factor is quite deciding. As a whole the Kosi river basin having the drainage texture between 0 to 6.26 meter. The texture of the different streams of the Kosi river basin has been calculated. There are variations in the texture from the plain part of the basin to the mountainous tract. The closeness is higher as the mountainous area and the stage of the river is late juvenile and thus the channels are widely spread. Thus the texture of Bhote Kosi river is likewise the texture of the other streams are found to occur 0.3, 0.6, 0.8 and 0.4 respectively which shows medium to lower texture.

CONSTANT CHANNEL MAINTENANCE: The inverse of drainage density gives the constant channel maintenance (Schumm, 1956, p. 607). This linear property is defined as area in square meter to maintain 1 meter of channel length.

Table - 5.16
CONSTANT CHANNEL MAINTENANCE

S.N.	Streams	Drainage Density	Length of Overland Flow	Constant of Channel Maintenance
		$Dd = \sum \frac{L}{A}$	$Lg = \frac{1}{2Dd}$	$C = \sum \frac{A}{L}$
01	Arun	0.82	0.61	1.23
02	Bhote Kosi	0.33	1.55	3.06
03	Tama Kosi	0.22	2.26	4.52
04	Dudhi Kosi	0.52	0.95	1.91
05	Sun Kosi	0.99	0.50	1.01
06	Tamur	0.70	0.71	1.42

07	Bagmati	0.15	3.36	6.73
08	Kamala Balan	0.39	1.29	2.58
09	Sapt Kosi	0.21	2.34	4.69
MEAN		**0.48**	**1.51**	**3.02**

Source: Computed from Topomap no. 72/G & 72K (for Indian part) and Google earth images (for China & Nepal part)

This quantitative expression is characterised of a drainage system. It shows (Table no. 5.15) the minimum limiting area required for the development of the channel, this value of channel is a measure of texture. It provides means of comparing the surface erodiblity and drainage network development. The constant channel maintenance of in case of Sapt Kosi river basin the value ranges from 1.01 to 6.73 and the average is 3.02.The value is indicative of heterogeneous climatic conditions of the region.

RELIEF ASPECTS: Relief aspects includes the analysis of the relationship between the area and altitude and the study of three dimension study of features of the basis involving area, volume and altitude of vertical dimension of landform where in different morphometric methods are used to analysis terrain characteristics which are the results of the basin processes (Fairbridge, 1968). Relief is set to be the range of the altitude (Smith, 1935). 'It is the difference between elevation and the given points' (Doore, 1951). 'Relief is related to the depth of dissection, the difference in altitudes' in which the former should be below and the later ago. 'The maximum basin relief is the elevation between the basin mouth and the highest point on the basin perimeter' (Strahler, 1968). Schumm, 1956 & Melton, 1968 defines the relative relief as the maximum relief H divide by the basin perimeter P while, Maxwell defines the relative reliefs as the maximum relief H divided to the basin diameter.

Relief measures are indicative of potential energy of a drainage system, intentionally of rate of erosion and transmiblity of rock debris. 'Relief and slope steepness together have a mark effect on the rates of surface runoff, infiltration and throughflow (Brunden, 1980).'

BASIN RELIEF: Basin relief, the elevation difference between head waters and outlet, is a measure of dividing force due to gravity acting on all flows on the basin. The basin relief (Table no. 5.17) is determining by subtracting the basin outlet elevation from the elevation of highest continuous contours of the head of the basin.

Table - 5.17
BASIN RELIEF

S.N.	Streams	Drainage Density	Length of Overland Flow	Constant of Channel Maintenance
01	Arun	8,848	146	8702
02	Bhote Kosi	8,012	1,128	6884
03	Tama Kosi	7,952	654	7298
04	Dudhi Kosi	6,820	543	6277
05	Sun Kosi	6,958	146	6812
06	Tamur	8,586	146	8440
07	Bagmati	2,650	55	2595
08	Kamala Balan	1,200	55	1145
09	Sapt Kosi	1,200	45	1155
			AVERAGE	**5479**

Source: Computed from Topomap no. 72/G & 72K (for Indian part) and Google earth images (for China & Nepal part)

The average basin relief is 5,479 meter. The highest value is 8,848 m and the lowest are 45 m the fairly high value of basin relief indicative of higher erosive activities and youthful stage particularly in the upper reaches of the basin.

RELIEF RATIO: Relief ratio (Rk) is the ratio of the maximum basin relief to the horizontal distance along the longest direction to the basin parallel to the principle drainage line. It is relatively the simple and effective index of basin gradient where the basin relief H is divided by the horizontal distance on which it is measured, the result is dimensionless relief ratio (Rk). 'Relief ratio is thus, the ratio of maximum basin relief to the horizontal distance along the largest dimension to the basin, parallel to the principle drainage line (Schumm, 1956).'

Table - 5.18
RELIEF RATIO

S.N.	Stream	Maximum Basin Relief (H)	Horizontal Distance (in km)	Relief Ratio
01	Arun	8,848	379	23.34
02	Bhote Kosi	8,012	140	57.23
03	Tama Kosi	7,952	81	98.17
04	Dudhi Kosi	6,820	170	40.12
05	Sun Kosi	6,958	270	25.77
06	Tamur	8,586	150	57.24
07	Bagmati	2,650	597	4.44
08	Kamala Balan	1,200	228	5.26
09	Sapt Kosi	1,200	350	3.43
			AVERAGE	**35.00**

Source: Computed from Topomap no. 72/G & 72K (for Indian part) and Google earth images (for China & Nepal part)

From the above table (5.18), the height relief ratio (Rk) exists in the river basin of Tama Kosi, 98.17. It is 57.23 of the Bhote Kosi, 40.12 in Dudhi Kosi, 57.24 in Tamur basin and gradually, relief ratio is found in decreasing order. It is lowest in the Sapt Kosi basin, it is 3.4.

The relief ratio can be used to rationalize basin dynamics (Chorley, 1950, p. 55) and the shape of the drainage basin. It is a bearing on the runoff of the stream. There is wide variation in the relief ratio. It is 3.43 to 98.17. It indicates quick runoff response to rainfall. The relief ratio exercises and important control over the sediment loss over the basin. The high variation in the relief ratio indicates towards the high loss of sediments as well as variations in the lithological characteristics.

ALTIMETRIC ANALYSIS: altimetric analysis helps in identifying and determining the erosion surfaces (Brush, 1961). H. Bauling (1935), has convinced by the soldiers on the valley sides due to successive lowering of the sea level, adopted an altimetric technique for the strategic. 'The altimetry represented by altimetric histograms, and curves represents frequencies of spot heights submits, highest points in grid squares and length of soldiers and Coles, spot-heights plus summit heights etc (Singh, S., 1985)'

Table No. 5.19
ALTIMETRIC ANALYSIS

S.N.	Spot Heights/ Summit's Name	Spot Heights (in m)	Cumulative of the Height
01	Mt. Everest	8,848	8,848
02	Kanchanjunga	8,586	17434
03	Mt. Malkalu	8,463	17049
04	Langtang Lirung	7,234	15697
05	Gauri-Shankar	7,145	14379
06	Shisha Pangma (Gosainthan)	8,013	15158
07	Abi	6,090	14103
08	Ama Dablam	6,814	12904
09	Baruntse	7,152	13966
10	Chamar	7,287	14439
11	Cho Oyu	8,188	15475
12	Gyachungkhang	6,960	15148
13	Hungchhi	7,029	13989
14	Pumori	7,138	14167
15	Ngo Jumba Kang	7,743	14881
16	Kabru I	7,353	15096
17	Ambulapcha La	5,845	13198
18	Kongma La	5,540	11385
19	Lapsang Bhanjya	5,161	10701
20	Minbo La	5,848	11009
21	Nangpa Bhanjyan	5,725	11573
22	Nire La	5,368	11093
23	Nup La	5,844	11212
24	Ragla Bhanjyang	4,900	10744
25	Sinelapche Bhan	4,600	9500
26	Thangla Bhanjya	4,820	9420
	AVERAGE	**6,681**	**13,176**

Source: Computed from Topomap no. 72/G & 72K (for Indian part) and Google earth images (for China & Nepal part)

From Figure no. 5.5 and Table no. 5.19 it is evident that there are wide variations in the heights as the uppermost catchment area presents a distinct height level where as it is different in the middle Himalaya, Shivaliks and the plain tract of Sapt Kosi river basin.

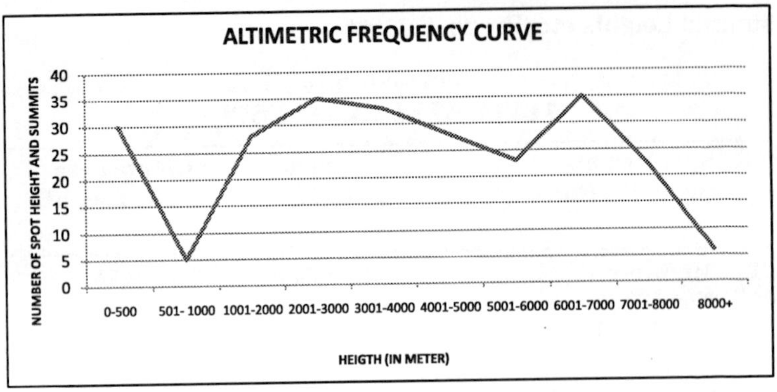

Fig. 5.5

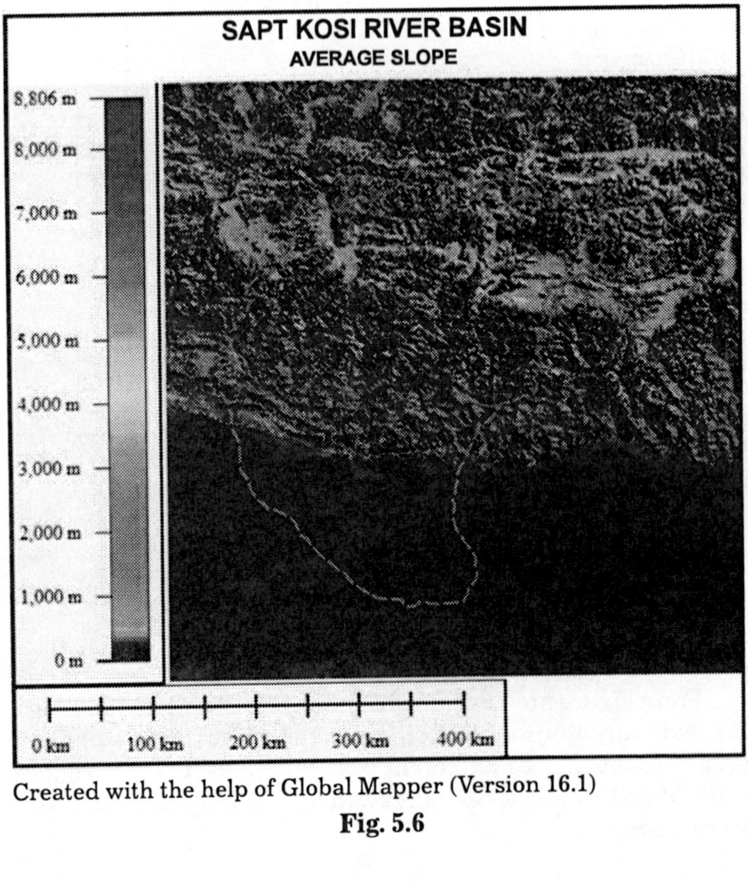

Created with the help of Global Mapper (Version 16.1)

Fig. 5.6

AVERAGE SLOPE: Slope is an angular inclination of terrain between hill top and valley bottoms, resulting from the combination from many causative factors like geological structure and frequency, dissection index. Computation of slope angle from topographical maps or through field measurement has remained a complex effort. It has been attempted by several geomorphologists like, J.L. Rich (1916); C.K. Wentworth (1930); G.H. Smith (1938 & 1939); W.C. Gleff (1950); A.N. Strahler (1956); O.M. Miller and C.H. Summersoo (1960); R.J. Eyles (1965). C.K. Wentworth method of widely used and accepted by the geomorphologists. The average slope of Sapt Kosi river basin has been computed using Global Mapper (version-16.1) data obtained from SRTM.cgiar website (90 m resolution).

RELATIVE RELIEF: Relative relief depicts the local relief (Melton, 1958B). Relative relief is an important morphometric variable used for the assessment of morphological characteristics of terrain and degree of dissection. Melton used relative relief Rhp expressed in percentage (%).

Table - 5.20
RELATIVE RELIEF

Stream	Maximum Basin Relief	Basin Perimeter	Relative Relief
	(in m)	(in km)	$Rhp = \frac{100(H)}{2580(p)}$
	(H)	(p)	(After Melton)
Arun	8,848	1,591	0.002
Bhote Kosi	8,012	525	0.591
Tama Kosi	7,952	369	0.835
Dudhi Kosi	6,820	384	0.688
Sun Kosi	6,958	783	0.345
Tamur	8,586	414	0.804
Bagmati	2,650	909	0.113
Kamala Balan	1,200	451	0.103
Sapt Kosi	1,200	656	0.071
Total/Mean	**5,803**	**6,025**	**0.395**

Source: Computed from Topomap no. 72/G & 72K (for Indian part) and Google earth images (for China & Nepal part)

The relative relief of the Tama Kosi is 0.835 which is height than the other basin. The relative reilief having more than 0.600 are Tamur (0.804), Dudhi Kosi (0.688), where as the other stream has relative relief of less than 0.500. It indicates that the streams having the height relative relief is more active in erosive activities and the stream with lower value particularly the stream of the plain tract are found to be comparatively less erosive.

RUGGEDNES NUMBER: Ruggedness is the product of local relief and drainage density. Ruggedness number use to measure slope steepness and stream length (Strahler, 1958). Ruggedness number (Hd) = Relief (H) x Drainage density (Dd).

Table - 5.21
RUGGEDNESS NUMBER

Stream	Maximum Basin Relief (in m)	Drainage Density	Ruggedness Number
(Rivers)	**(H)**	**(Dd)**	**(Hd)**
Arun	8,848	0.82	7255.36
Bhote Kosi	8,012	0.33	2643.96
Tama Kosi	7,952	0.22	1749.44
Dudhi Kosi	6,820	0.52	3546.4
Sun Kosi	6,958	0.99	6888.42
Tamur	8,586	0.70	6010.2
Bagmati	2,650	0.15	397.5
Kamala Balan	1,200	0.39	468
Sapt Kosi	1,200	0.21	252
Total/Mean	**5,803**	**0.48**	**3245.7**

Source: Computed from Topomap no. 72/G & 72K (for Indian part) and Google earth images (for China & Nepal part)

The ruggedness number computed in Table no. 4.21 suggests the slope intensity. Increase in drainage density indicates towards the constant relief, the average horizontal distance divide to adjacent channel is reduced, with in accompanying in slope steepness. The ruggedness number of Sapt Kosi river basin as a whole is 3,245.7 which indicate high steepness. It is 6,888.42 to Sun Kosi, 7,255.36 of Arun, likewise 6,010.2 to Tamur and 2643.96 to Bhuti Kosi. Number is comparatively lesser to the other streams. The high value of ruggedness number and low drainage density of the Kosi river as whole indicate higher relief value. The low relief of ruggedness number 252 to Sapt Kosi and 468 to Kamla Balan and higher drainage density indicates lower relief. High value of ruggedness number also indicates that the slopes are steep and long.

DISSECTION OR AERATION OF SAPT KOSI RIVER: The dissection or aeration of a landscape means the extent to which has been opened up by erosion (Clark, 1966). Dissection index is the ratio of the maximum relative relief to the maximum absolute relief and is indicator of the nature and dissection of terrain (Slaucitajs, 1936, p.p. 216-240).

Table - 5.22
DISSECTION INDEX

Index Group	Area (in km)	Percentage (%) of the Total
0-900	32953	36.06
900-1350	4843	5.28
1350-1800	5488	6
1800-3000	5108	5.57
3001-4500	22707	24.82
4500-6000	16601	18.15
6001-7500	2685	2.93
7500+	1091	1.19
TOTAL	**91476**	**100**

Source: Computed from NATMO (Physiography of India) Computed from using IGIS (verstion 1.1)

The above table potraise the following facts.

i) The lower catchment area of Kosi river basin have low dissection index. It constitutes 36.06 % of the total catchment area. This morphological unit generally corresponds to the lower most part of Bihar plain and to some extent the Terai region of Sapt Kosi river basin. Generally, the low dissection

Table - 5.23
HYPSOMETRIC INTEGRALS

Contour	Absolute Ratio	Absolute Ratio	Percentage of	Percentage of	Cumulative	Cumulative
in meter	$(\frac{h}{H})$	$(\frac{a}{A})$	$(\frac{h}{H})$	$(\frac{a}{A})$	$(\frac{h}{H})\%$	$(\frac{a}{A})\%$
0-900	0.67	36.06	10.47	36.06	10.47	36.06
900-1350	0.81	5.28	12.66	5.28	23.13	41.34
1350-1800	0.84	6	13.12	6	36.25	47.34
1800-3000	0.8	5.57	12.5	5.57	48.75	52.91
3001-4500	0.78	24.82	12.19	24.82	60.94	77.73
4500-6000	0.77	18.15	12.03	18.15	72.97	95.88
6001-7500	0.85	2.93	13.28	2.93	86.25	98.81
7500+	0.88	1.19	13.75	1.19	100	100
TOTAL	**6.4**	**100**	**100**	**100**	-	-

Source: Computed from NATMO (Physiography of India) Computed from using IGIS (verstion 1.1) index corresponds with Tame and subdued relief.

ii) Six or seven pockets in the Shivalik and middle Himalaya also correspond to low dissection index of 5.28 %. It is primarily due to the low dissection index. The area is valley and constitutes almost mountainous and hilly land.

iii) It is also depicted by the table that the low dissection index generally, corresponds with the areas of low relative relief. The explanation of this is forced in the Terai and the Shivalik Himalaya of the basin.

iv) The areas of medium dissection index are extensively existing the lower Shivalik, the Terai region and as well as to same extend the middle Himalaya.

v) The regions with high dissection index which cover 4.12 % of the total catchment area and generally found to occur in the high peak, mountain and high altitudinal areas.

Dissection index from 1.19% to 24.82% has greater area in the northern hilly region and particularly in the middle and the upper catchment area. Incidentally, all this areas of high dissection index are the high altitudinal peak areas of the mountains.

HYPSOMETRIC INTEGRAL: 'Hypsometric is the measurement of the interrelationship of area and altitude (King, 1966).' These values are plotted into graph hypsometric curve results.

AREA-HEIGHT CURVE: The area height curve represents the absolute or relative areas of land between the adjacent contours. In this curve Fig. No. 5.7 area calculated in terms of percentage and ploted against height, which are the absolute valleys. One mode at average 150 meter above sea level and other between 1,500 to 8,000 meter. Through this area-height diagram relationship between the successive contours and their respective heights is determine.

Table - 5.24
AREA HEIGHT CURVE

S.N.	Height	Area	Percentage of the total
	(in meter)	(in Sq. km.)	(in %)
01	0-900	32953	36.06
02	900-1350	4843	5.28
03	1350-1800	5488	6
04	1800-3000	5108	5.57
05	3001-4500	22707	24.82
06	4500-6000	16601	18.15
07	6001-7500	2685	2.93
08	7500+	1091	1.19
TOTAL		**91476**	**100**

Source: Computed from NATMO (Physiography of India) Computed from using IGIS (verstion 1.1)

PERCENTAGE HYPSOMETRIC CURVE: Percentage hypsometric curve involves two ratios of relative height ($\frac{h}{H}$) denotes height between two successive contours and H indicates total heights, plotted on the ordinate and relative area ($\frac{a}{A}$) in which a denotes area between two successive contours and A

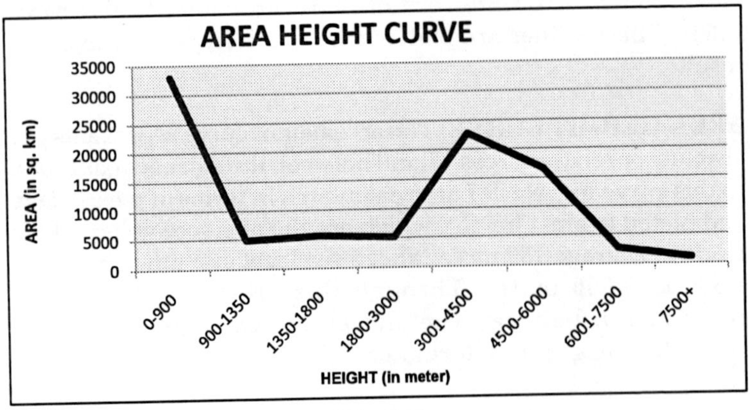

Fig. 5.7

indicates the total basinal area plotted in the abscissa in terms of percentage.

Table - 5.25
PERCENTAGE HYPSOMETRIC CURVE

S.N.	Height (in meter)	Area (in Sq. km.)	% of the Area (in %)	Cumulative Area in % (in %)
01	0-900	32953	36.06	36.06
02	900-1350	4843	5.28	41.34
03	1350-1800	5488	6	47.34
04	1800-3000	5108	5.57	52.91
05	3001-4500	22707	24.82	77.73
06	4500-6000	16601	18.15	95.88
07	6001-7500	2685	2.93	98.81
08	7500+	1091	1.19	100
TOTAL		**91476**	**100**	-

Source: Computed from NATMO (Physiography of India) Computed from using IGIS (verstion 1.1)

MORPHOMETRY THROUGH PROFILE: The vertical direction of the terrain forms are visualized by contour lines. Profile are drawn from the contour may are great assistance. Profiles are best suited for the analysis of relief and surfaces of terrain because it provides a visual perception of the actual nature of the terrain. 'The mark of contour lines for relief representation lies not only in their being a systematic and geologically satisfactory procedure, but even more important that the result in a good visualization of the natural terrain by way of their coherent pattern' (Toepter, 1964, p.15). For visualization of relief all over the basins two types of profile have been done. i) River profile, and ii) Terrain profile. For the analysis of river morphology two types of profile have been drawn. i) Longitudinal, and ii) Transverse profile, for the visualization of the terrain of the whole basin.

a) Terrain profile have been drawn as 1. Serial profile, 2.Superimposed profile, 3. Projected profile and 4. Composite profiles.

Five longitudinal profiles A-B, C-D, E-F, G-H, and I-J and

MORPHOMETRY THROUGH PROFILES

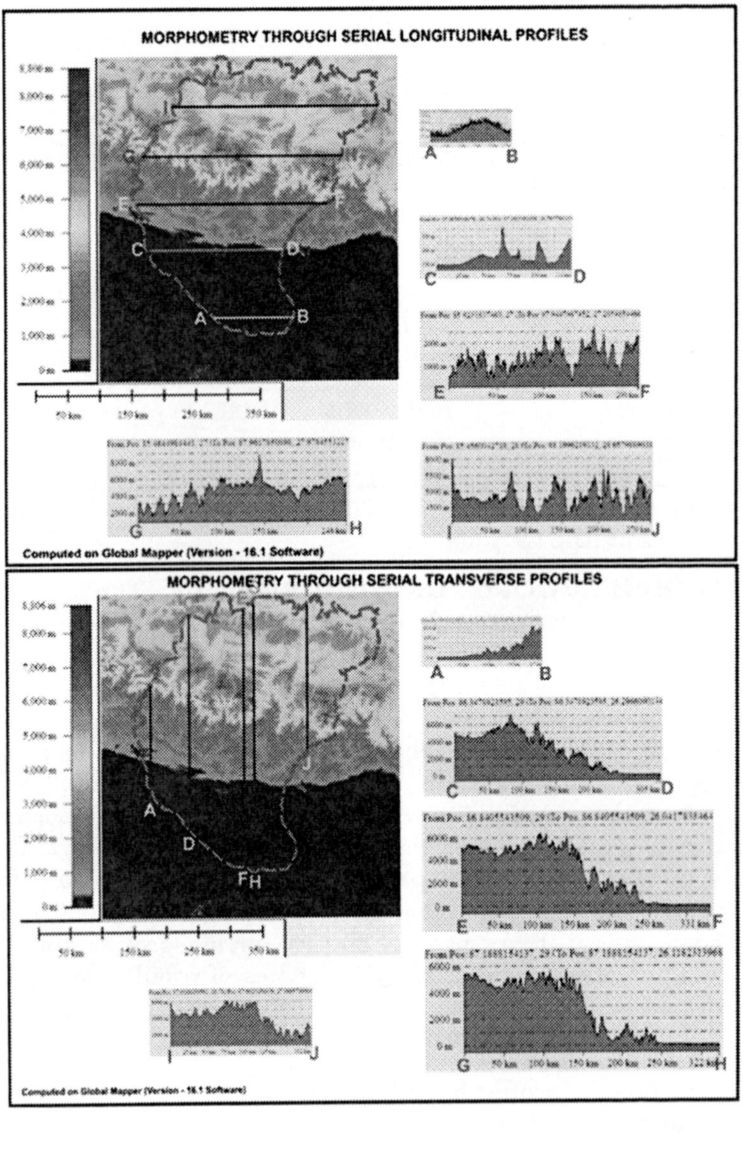

Fig. 5.8

PROFILE ANALYSIS

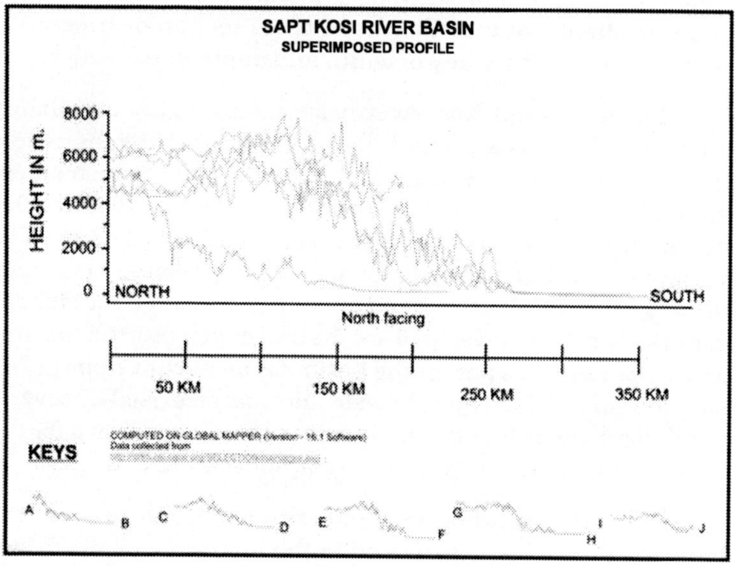

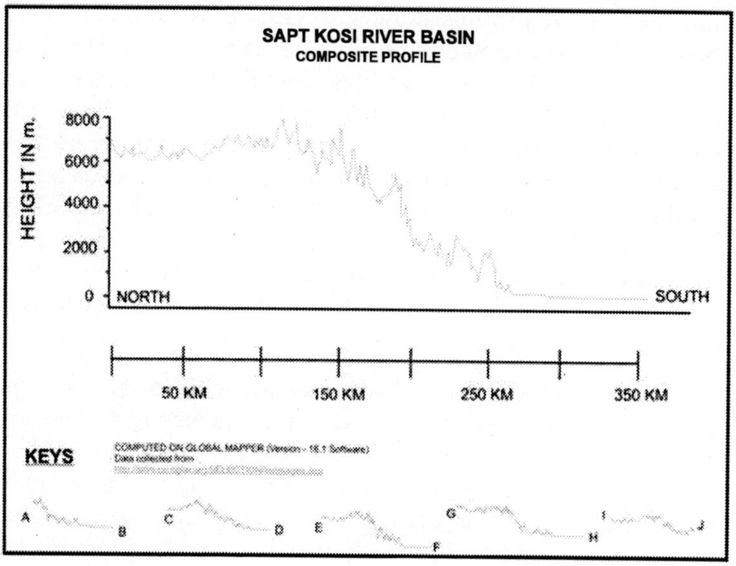

Fig. 5.9

five transverse profiles K-L, M-N, O-P, Q-R and S-T has been drawn at various places. Longitudinal profiles which is also called the longitudinal sections involving the study of channel slope gradient. On the other hand cross section or transverse profile involves the study of width and depth of the valley.

The whole Sapt Kosi river basin forms highly undulating and altitudinal area. The hilly tracts constitute the whole northern part while the plain tract occurs in the lower catchment area. The hills and ridges are of varying elevation. 3,000 m 6,000 m and the intra scarp valleys of varying width and depth. The variation in altitude and extension in largely related to tectonic nasty, geological structure and erosion process. The scarps are commonly ridges and sparse are the common feature of the hilly area. The northern part of the basin has an abrupt slope of 20^0-90^0. The surface is rugged, broken and characterised by several parallel hills. The longitudinal profiles have been drawn for the visualization of the major physiographic features.

Under this heading morphometric through profiles the landscape have been discussed in the investigation of surface morphology the profiles show with great precision the form of the land. The whole Sapt Kosi river basin has been divided into five zones of which serial profiles have been drawn across the grain of the country from west to east and viewed from the south. The serial profiles shows the high undulation of surface morphology in one line across the country, superimposed profile exhibit the morphological unity it any and existence of the certain level or erosional or structural surfaces. Profiles of different types reflect the corresponding slopes, periods of peniplaination and of aggradation which reveals significant surface forms.

The projected profiles drawn to shown discordant summit here and the composite profile exhibits surfaces ruggedness and the sky line of the landscape. It is also very much suitable to exhibit the panoramic view of the landscape as well as the main water divides. The present profiles has been drawn longitudinally and transversely in which formed exhibit the growth and evolution of the main streams where as the secondary one exhibit the present erosion level of the basin area. From the Fig. no. 4.8 the following points can deduced.

i. Five levels are erosional are clearly marked in the Kosi catchment area. It is at an elevation of a) 3050 to 4500 meter, b) 1870 to 3000 meter, c) 1350 to 1800 meter, d) 950 to 1350 meter and e) 600 to 900 meter.

ii. The average elevation of the entire basin is unusually, very high (5,803 m). It is between 5000 m to 6000 m above sea level. This basin is a part of Himalayan region and gargatic drainage system.

iii. The northern and north western part of catchment area is relatively very high. Sapt Kosi main stream rises from eastern fore front of the Everest and the lower most part of the basin is relatively very low with an elevation of 150 meter from msl.

iv. The landscape of the basin shows very panoramic as the basin water shed area basin area is full of peak of different elevations and almost all the peaks are covered with snow.

v. The composite profile reveals the fact that the landscape has been dissected by numerous streams. The interfluves are high but not extensive. The interfluves are dissected by numerous hilly streams.

vi. The general slope of the basin is towards west to east in the upper Himalaya and west to south which again turns towards south.

vii. The region is highly dissected highland in the upper reaches and peaks are truly conical

viii. The gradient of the basin is from north to south. The nine tributaries works as the local base level. Several adventeceous streams gets theirs origin in the Terai region and the watershed area of the different streams

ix. The whole catchment area is dominated by high fluctuations and variations in the slope and peaks of different elevations.

x. The interfluves are very high extensive with the high degree of slope numerous small hilly streams have dissected the surface considerably.

xi. The higher value of the relative reliefs of the interfluves indicates the juvenile and initial stage of the cycle of erosion.

CHAPTER – 6

CHANNEL MORPHOLOGY

It is with the "form", "materials" and "processes" of the earth's surface that 'geomorpholgy is concerned. Under the domain of such definition of 'geomorphology' the term 'river channel' refers to the 'trough like form', which contains a 'river' and is shaped by the force of water flowing along it (Whittow, 1984). Channels generally transmit water and materials from the drainage basin in an effort to quantity erosion. They normally carry an excessive amount of sediment washed off the valleysides or acquired from the channel floor and banks. 'Flood plain' is that part of river valley, adjacent to the channel over which river flows in times of floods (Whittow, 1984).' Actually it is a zone of low relief and gentle slope gradient. That may incorporate oxbow lakes, point bars, abundant channels and scrowlls etc, all indicating the fact that 'river channel' has shifted its position continuously during past and present regime of the stream. The flood plain generally composed of alluvium which buries of the rock floor of the valley to variable depths. The alluvium deposits of the flood plain would have been formed either within the channel itself (C.75%) of the alluvium or as over bank deposits (C.25%).

Channel morphology or river channel morphology includes the consideration of

i) Channel geometry or channel cross sectional characteristics (that is channel length, channel depth, channel width, channel wetted parameters, channel bends/meanders, channel slope etc.).

ii) Hydraulic geometry: Channel fluid dynamics a) discharge of the river, b) velocity of the channel

iii) Channel types

iv) Channel bed topography or bed configuration

v) Channel patterns

The channel morphology of Kosi river basin has been dealt under the above heads.

CHANNEL GEOMETRY: 'The fluvial system is powered by the conversion of potential energy of solar distillation and gravity to the kinetic energy of motion and heat. Most of the energy is lost friction of internal turbulence of flowing water, perhaps, 2-4% of the potential energy of the water flowing down hill over and erodible bed is connected to mechanical work of erosion and transportation which becomes tools for an effective channel geometry (Rubey, 2008).'

Channel geometry refers the size and shape of the cross sectioned and longitudinal channel form and includes channel width, channel depth, wetted parameter, channel slope, channel bends, shape of the channel thalweg and their interrelationships (Singh, 2006; Reprint).' A river channel represent water course of a river confined within the valley walls or both the sides. 'Infect, river channel presents three dimensional form-define by its slope, cross section and pattern (Petts and Foster, 1985).'

The channel geometry which tends to the channel width, channel depth, wetted parameter, and channel bends, shape of channel thalweg has been analyse here by of Sapt Kosi river

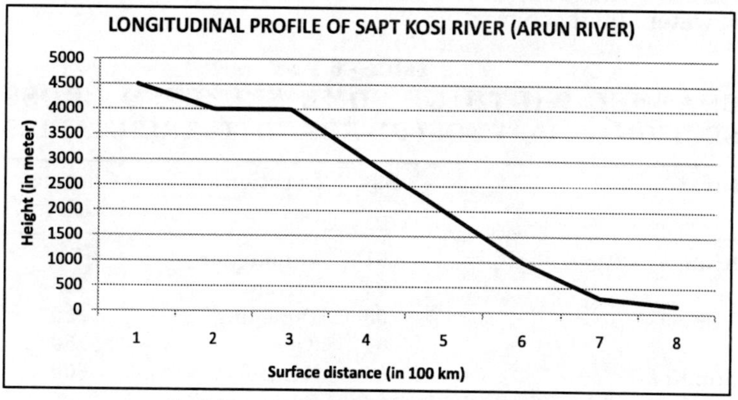

Fig. 6.1

basin. The channel geometry is the mechanism of a river channel a number of work has been done in this regards in which the important are Leopold at. el., 1964; Leliacasky, 1966; Morisawa, 1968; Raudkivi, 1967; Chorley, 1969a; Fraf, 1971; Schumm, 1972, 1977; Gregory and Walling, 1973; Rechards, 1982; Hey, et al., 1982.

CHANNEL LENGTH: Channel length is the distance of channel from source to mouth of a river and connects all the mid points the channel, channel width. The length of Sapt Kosi river basin has been calculated by law of stream length. 'The proportion of increase of mean length of stream of two successive basin order is defined as ratio (R_L).' The best description of river's long profile is its slope or gradient, which is the graph representation of ratio of full of the channel to its length over a given reach. It is thus bottom in longitudinal view.

The total stream length of stream order is inversely related to stream order that i.e. total stream length decreases from the lower order to the successive higher order. It has been calculated for Sapt Kosi river as the stream order increases the stream mean basin area of each stream order also increases. On the basis of above formula μ of the basin has been calculated and found 7.79 mean.

CHANNEL WIDTH: Channel width (W) at any given point along the course of a river represents straight cross sectional distance of a channel representing stage of a river. It is thus represent that channel width varies with changing in volume of water and discharge.

Table - 6.1

CHANNEL WIDTH OF SOME SELECTED CROSS SECTION OF SAPT KOSI RIVER AND ITS TRIBUTARIES

River		S.N.	Location	Channel width (m)
Arun		01	Tumlingtar	70
		02	Triveni	255
Tamur		03	Taplejung	305
		04	Kabeli	150
Sun Kosi		05	Ramechap	600
	Bhote Kosi	06	Pachuwarghat	70

	07	Brahmabise	500
	08	Kodari	23
	09	Tatopani	30
Tama Kosi	10	Singati	140
	11	Charikot	150
Dudhi Kosi	12	Namche bazaar	90
Balepi Khola	13	Balepi	300
Kamla Balan	14	Jhanjharpur	296
	15	Siraha	40
Bagmati	16	Kathmandu	15
	17	Pipraha Ghat	12
	18	Darbhanga	500

The channel width considerably changes in the river humly seasonal regime. In case of cross sectional shape of the channel in which is irregular, channel width increases with increasing discharge and volume of water nearly high gradient zone. The different tributaries with its main flow valley deep increases

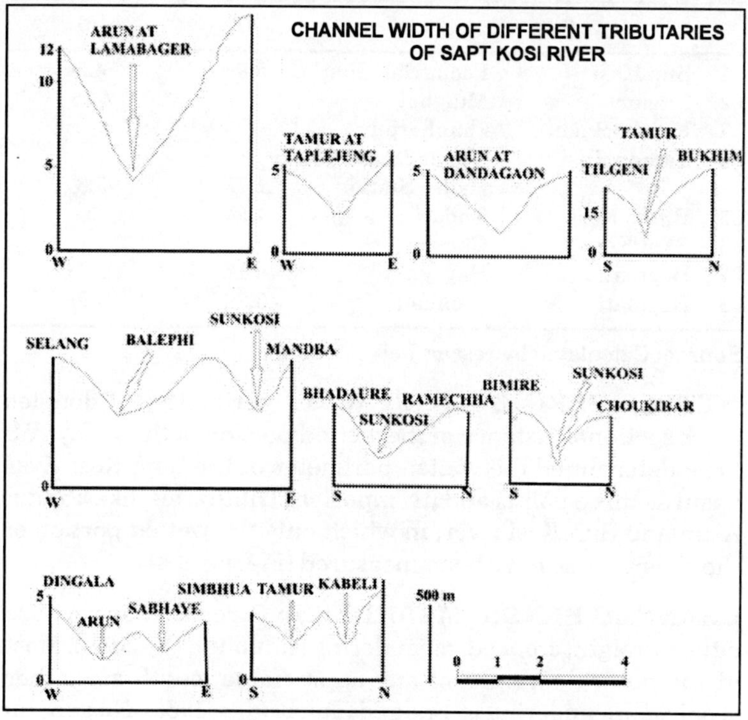

Fig. 6.2

and gorge type and width is lomted compared to depth where as in the middle zone when river start flowing east to west the valley becomes comparatively wide except in areas where rocks are hard dolomitic, quartzite and granitic.

CHANNEL DEPTH: The depth of channel denotes vertical distance from the water level in the channel to the channel bed. Maximum channel depth (Table no. 6.2) is measured from the level of water of bankful stage of the river, to the lowest point of the channel bed (Fig. No.: 6.2). while the mean channel depth (ð) denote average of channel depth lowerR at different stages of channel.

Table - 6.2

CHANNEL DEPTH OF SOME SELECTED POINTS

S.N.	Name of Stream Channel	Depth Station	Channel Depth	Mean Channel Depth
			(d)	(ð)
01	Sun Kosi	Pachuwar Ghat	5.8	4.2
02	Tmaur	Mulghat	7	4.25
03	Kamla-Balan	Jhanharpur		
04	Tamba Kosi	Upper Tamba Kosi Hydro Station	4.7	3.8
05	Bhote Kosi	Kodari	4.5	3
06	Sapt Kosi	Chatra		
07	Bagmati	Hayaghat	49.17	42.41
08	Bagmati	Benibad	48.76	34.96

Source: Calculated by researcher

WETTED PERIMETER (P): Wetted perimeter (p) denotes cross sectional distance of the wetted portion of the valley. We have determined the wetted perimeter of the Sapt Kosi river basin as three points and its important tributaries, like Tamur, Arun and Sun Kosi river; in which only the wetted portion of the river course have been measured (Fig. no. 6.4)

CHANNEL BENDS (MEANDERS): Stream channels are either straight, crooked, meandering (sinuous or braided). Most of the natural streams of any appreciable length are either crooked, meandering or braided (Fairbridge, 1968). No general theory of meander is at satisfactory, however, infect present evidence suggest that no single theory well explain the formation and characteristics of all meanders and that few of the physical principles involve have yet been clearly identified. The consensus of opinion among the Geomorphologists and Hydrologists is that the effect Helical flow of water which is due to combined effect of frictional and centrifugal forces is the dominant factor in the meander development. Ledivasky, (1955) Prus-Chascluski, (1954) are the main exponents of the above view. Matthes (1941, p.p. 644), attributes the development of point bars, Quraishy (1944) to 'Skewshoals' and Warner (1951) to 'some local impulse or disturbance as the cause of traverse oscillation in straight channels and initiations of meanders. Whatever may be the causes of meandering, it is a truism that

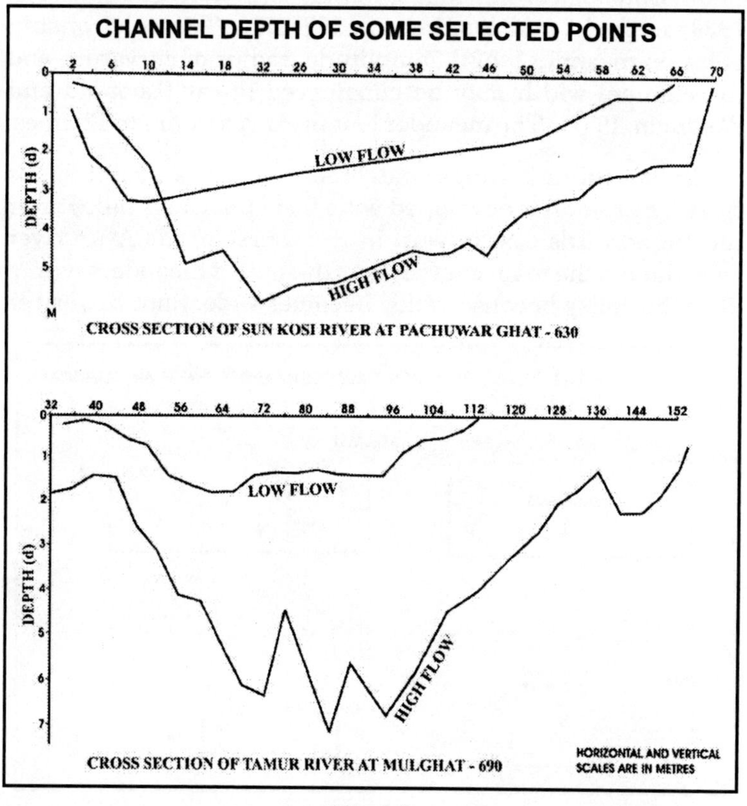

Fig. 6.3

river follows a winding course and in nature starlight channel are so rare, 'as to be almost non-existant (Leopold and Woldman, 1957).

In Sapt Kosi river basin all the streams show some amount of sinuosity. The measurement has been taken only at seven big and medium size river of the basin that is the Kosi main stream (Sapt Kosi), Sun Kosi, Tamur, Indrawaty, Bhote Kosi, Kamla-Balan and Bagmati. Riffles, point bars and pools are the common in the meandering reaches of the rivers, meandering channels are characterised by deep at the concave bank where the water surface is super elevated. All the meandering channels are quasi-equilibrium stage. The statistical and quantitative study of geometry of meander has been done by the Geomorphologists, like Inglis, 1937-1949; Jefferson, 1902; Bets, 1939; Leopold and Woldman, 1957-1960. The relationship between meander length, amplitude, radius of curvature and the channel width may be considered linear (Leopold and Woldman, 1957).' The meander length ranges from 7 to 12 times.

Sapt Kosi in its upper catchment area is a mountainous river. The river has developed wide flood plain and valley after turning towards east to west in the course of the Arun river even though the main course is north-south it meanders within limit the valley because valley becomes wide. Sapt Kosi river

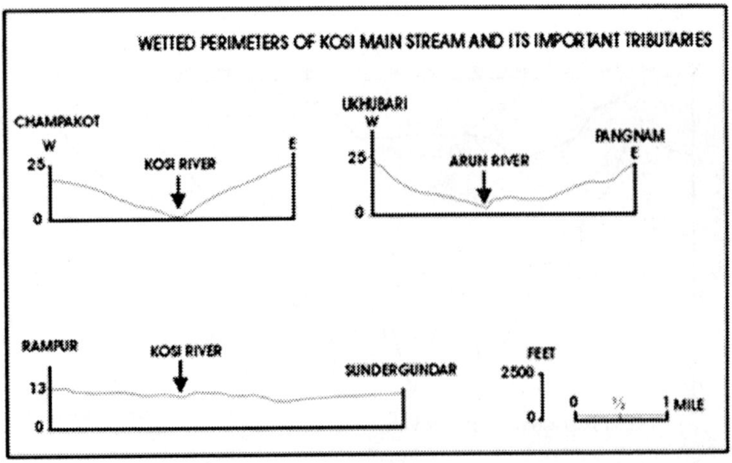

Fig. 6.4

Table - 6.3

GEOMETRY OF MEANDERS AT DIFFERENT CHANNEL SEGMENTS OF SOME STREAMS OF KOSI RIVER

S.N.	River	Location	Location	Meander Length	Meander Width	Area	Dist-ance	Mean Radius	Sinu-osity
			(LAT-LONG)	(Km)	(Wm)	In sq. km	(m)	(rm)	
01	SAPT KOSI	North-East of Khagaria	26°20'94"N/85°91'41"E	2.57	3.23	13	6.6	1.28	1.15
02	SAPT KOSI	Dhobinia milik	25°28'08"N & 87°13'35"E	3.4	2.1	12	5	1.23	1.16
03	SAPT KOSI	Partapnagar	25°27'21"N & 87°04'57"E	2.1	3.5	15	7.3	1.4	1.09
04	BAGMATI	Kalyanpatti	26°18'46"N/85°52'16"E	2.6	2.4	12	5.4	1.2	0.91
05	BHOTE KOSI	Dolalghat	85°39'37"N/85°43'17"E	1.2	0.4	7	3	0.8	0.012
06	SUN KOSI	Ramechap	27°17'10"N/86°14'44"E	2.03	05	6	5	0.5	0.011

Source: Computed from IGIS (version 1.1) and topo maps for Indian part.

(Indrawaty) starts meandering after the Dolaghat where the river is flowing near one bank and deflected to other bank in the next few kilometers. Most of the river like Kamla Balan, Bagmati in Terai after reaching mid part starts meandering heavily; most of the meanders are seen in Indo-Nepal border. River changes their courses with the result of number of oxbow lakes are found. In the high flow the channel which is 120 m and max depth is 7.3 meters covering a cross sectional area 2171 sq. mile with discharge of 1860 cusec and velocity was 3.95 m/sec. whereas, the same river at low flow has channel .83 m wide and the maximum depth is 1.8 meter. Velocity is 0.64 m/Sec and discharge 54.5 cusec. Sun Kosi at Pachuwarghat has 70 meter wide river channel, being rocky the channel is not wide and hence the main variation is the scaring depth. In high flow V= 3.06 m/sec had discharge 6.73 cusec whereas, in the low flow 72.9 cusec with velocity 0.65m/sec and depth was 6 meter.

Later on the same place the channel is deepened to 10 meters and velocity increased 4.8-8 m /sec to cause the flow 2,000 cusec and cross sectional area 498.40 sq. m and low flow the cross section area of channel was 101.67 sq. m and velocity 0.43 m/sec and discharge was 43.68 cusec. This indicates that each time nearly 6 meters column of bed load is removed at the onset of flood to accommodate the flow and deposited the same slowly in the 10 m flow.

CHANNEL SLOPE: The channel slope or gradient in downstream direction is called channel gradient or slope. 'Different long profile related to long term gradient development, influence by tectonic history, base level change and the climate change. The degree of concavity increases with relief but accessively concave profiles are associated with river, which has extended over quaternary estuarine sediments at sea level retreated by flow diversion, under-concavity relates to lithological controls run off and sediment supply. Concavity tends to be more accentuated if the downstream increase of discharge is rapid and to approach of linearly where discharge increases slowly, such as within permeable catchment (G. Petts and I. Foster, 1985, pp. 104-110).

Sapt Kosi river basin is marked by heterogeneous landform

Table - 6.4

GRADIENT OF SOME SELECTED POINTS

River	S.N.	From	To	Distance	Gradient
		(height in m)	(height in m)	(in km)	(in meter per km)
Arun	01	Tibet (3048)	Naktang (2319.5)	9.6	121.3
	02	Naktang	Sedua (1243.6)	58	29.8 (deep gorges)
	03	Sedua	Maleng (762)	11	68.6
	04	Maleng	Ramche (609.6)	22.5	10.7 (wide valley with sandbars)
	05	Ramche	Mughghat (457.2)	48	4.9
			Total/Average	**149.1**	**47**
Tamur	06	Sakathum (1524)	Khokling (762)	25.7	47.5
	07	Khokling	Angbung (609.6)	19.3	12.5
	08	Angbung	Mahasar (457.2)	22.5	10.7
			Total/Average	**115.5**	**23.6**
Sun Kosi	09	Kodari (1524)	Chakhukhola (1371.6)	7.7	31.7
	10	Chakhukhola	Ghumtang (1219.6)	9.6	25.3
	11	Ghumtang	Gurdum (1066.8)	16	15.2
	12	Gurdum	Tokarbu (914.4)	9.6	25.3 (rapids)
	13	Tokarbu	Bhuka (762)	18	13.7 (Sandbars)
	14	Chuka	Pankhar (609.6)	19	12.5 E.W.
	15	Pankhar	Benighat (564.5)	16	4.6 E.W.
	16	Benighat	Suruug Kodule (457.2)	61	2.7

				Total/Average	**156.9**	**16.4**
Tamba Kosi	Khimti Khola	17	Lomsa (3048)	Purnagoan (1524)	19	126.8
		18	Purnagoan	Arkhaule (762)	13	95
		19	Arkhaule	Benighat (564.5)	25.7	12
				Total/Average	**57.7**	**77.9**
	Likhu	20	Ranganag (7500')	Ekgaon (1524)	13	95
		21	Ekgaon	Gothpani (762)	22.5	54.3
		22	Gothapani	Dihi (609.6)	12.8	19
		23	Dihi	Kolanger (518.2)	6.4	23
				Total/Average	**54.7**	**47.8**
Dudhi Kosi		24	Jubing (1524)	Chisankhu ghat (762)	38.6	31.7
		25	Chisanku	Kharpa (609.6)	6.4	38
		26	Kharpa	Kudule (472.44)	29	7.6
				Total/Average	**74**	**25.7**

Source: Calculated from toposheet no. 72G & 72K for Indian part and Google earth image and field excursion for Nepal & China part.

characterised by more than 62% of the catchment area is highly undulating with high gradient. The high gradient leads to a distinct landform and topographic development in a basin area. Sapt Kosi river basin from the higher Himalaya to the Shivalik Himalayan ranges from more than 90^0 angle of slope gradient.

HYDRAULIC GEOMETRY OR CHANNEL FLUID DYNAMICS: channel morphology is the result of mutual interaction of four broad categories of variables viz, a) Fluid dynamics, b) channel character or channel configuration, c) sediment load and d) bed and bank materials.

FLUID DYNAMICS: Discharge is measured by multiplying cross sectional area at which gauging station of average velocity of water current.

Discharge (Q) = w x d x v

Where, w = channel width

d = channel depth,

v = average velocity

Overall, geometry of a river channel is control by independent variable of discharge and load i.e. climate and geology of a watershed (Morisawa, 1985). Channel configuration, sediment load, bed and bank materials all depends on the discharge, velocity of the current. Morisawa has observed that 'adjustment mechanism include erosion or deposition in channel to change its form, shape or pattern and creation and movement of bed form. Scour of the bed will result in a lower gradient increase depth of catchment and decreased water velocity. Scour on the bank will increased the channel width and decrease velocity and depth with given discharge' (M. Morisawa, 1985).

Variations in channel morphology as a power function of discharge. The discharge of Sapt Kosi river basin has been calculated in view of the above equation.

SEDIMENT LOAD: sediment load of Sapt Kosi river basin has been assessed in two forms a) suspended sediment load and b) bed material load. Sediment load is an important variable which influences channel forms and hydraulic geometry. Suspended sediment load is kept in suspension as wash load in the water of river channel and its transport downstream depends on turbulence and channel velocity, where as the movement of bed load depends on the shear on the bottom of the channel. The Kosi river basin characterised by large size materials of high caliber; It is particularly in Tibet and plain tract of India as the channel of Sapt Kosi river in the upper catchment area is highly turbulent and velocity characterised by greater width-depth ratio in the lower catchment area and narrower in the upper catchment area. 'Larger the sediment load and larger the size of particles that must be transported through a stream, the steeper is gradient of the channel is required to transport this material (Chorley, 1985).' suspended sediment load is kept in suspension as wash load in the water in the river channel and its transport in downstream depends on turbulence and channel velocity, where as the movement of bedload depends on shear on the bottom of the channel. According to Leopold and Moddock, 1953, suspended sediment load increases more rapidly than water discharge at a station that is at a gauging station. 'suspended load is the material moving in suspension in the fluid, being kept in suspension by the turbulent fluctuation as per the theory of suspended sediment distribution (Garde, 2006).'

BED MATERIAL LOAD: bed material load especially in stream is difficult to measure, it is clubbed with contact load and sediment moved on or near the bed is the bed material load. The high bed load characterised by large size material of high caliber, is transported in a wide and shallow channels because, 'A wide, shallow channel increases turbulence and the possibility of hydraulic lift as well as increased bed shear (Morisawa, 1985).' Thus, it is clear that the channels, which transport large quantity of bed material load are characterised by greater width-depth ratio, while channel which transported small quantity of bed material load, have small width-depth ratio. Thus, 'the larger the sediment load and larger the size of the particles that must transported through a stream the steeper is the gradient of the channel is required to transport this material (Ackers and White, 1980).'

Sapt Kosi river basin is characterised by the longer sediment load and larger size of transport particles which are being deposited in the Terai region. The whole Terai region is an area of high bed material load. A huge quantity of bed material load being transported by the rivers but the same conditions is not prevails in the upper reaches where a narrow speedy channel transport comparatively less material.

CHANNEL TYPES: Broadly the channels are divided in two categories a) Bed rock channel, b) Alluvial channel.

BED ROCK CHANNEL: Bed rock channel also referred as rock channels. Bed rock channels occur where ever potential role of removal exceed sediment supply. It is mainly found in the high mountain areas where steep slopes, glaciated hard rock regions and in areas under going active tectonic upliftment (Petts and Foster, 1985, pp. 85-90). There is a deep cut river bed and wide valley representing example of canyon. Sapt Kosi river basin is example of both types of channels i.e. the rock channel as well as alluvial channels. It is in the upper catchment area represents an example of rock bed channel where as in the lower catchment area represents alluvium channels.

ALLUVIAL CHANNELS: This type of channel develops in the region of sedimentation or alluvium where the deposit of sediment mostly fluvial origin has taken place. Sapt Kosi river basin in the Gangatic plain in the lower catchment area where thick deposits of sediment of mostly fluvial origin are found. Alluvial channels are also characterised by sinuous to meandering and braided channel pattern, depositional and erosional bed topographic features as sandbars, pools and riffles, sandbar islands and dunes etc.

The alluvial channels are highly migratory. It is because these channel deposited material which are highly erodible; enhance, these charge the form and pattern with time. Sapt Kosi river is world's notorious and highly migratory channel. First, it is depositional and second it eroded away the deposited material very rapidly.

From the stand point of degree of temporal changes alluvial channels are categories into two types. i) Stable alluvial channels, ii) Unstable alluvial channels. Straight channels are stable where as braided channel are unstable. On the basis of variables which influence and control on channel morphology alluvial channels are classified into five categories which are follows:

i) Suspended load straight channel: characterised with straight course and uniform depth.

ii) Mixed-load-straight channel: it is mostly the small course sediment with sinuous thalweg.

iii) Suspended load channel: with high sinuosity uniform channel width stable banks.

iv) Meander braided transition channel: it is characterised by large sediment load having greater proportions of sand, gravels, cobles wide and shallow channels, variable channel width, chute-cutoff and shift in thalweg and meander.

v) Bed load channel: also referred as bar braided stream. It is characterised by unstable creliton, large sediment load fairly larger size and course in texture erodible bank materials etc.

Sapt Kosi river basin in its lower catchment area is the example of mixed load channel, suspended load channel as well as bed load channels.

CHANNEL BED TOPOGRAPHY OR BED CONFIGURATION: Channel bed topography refers to configuration of the river bed in terms of positive or negative features i.e. presence or absence of riffles or pools, sand bars and sand islands, shoals, sand dunes etc. Profile of a river is characterised by irregular course having a series of alternating sage and gentle segment where in steep segment denotes rifles while gentle segment call pool.

'A pool is characterised by a water surface profile less than the mean stream gradient and by finer bed material, where as riffle has a water surface slope steeper than the near stream gradient and is compared of coarse bed material (Morisawa, 1985).' 'A riffle is a topographic high area produced by lobate accumulation of relatively coarse sediment and a pool is a topographic low usually characterised by finer material (Petts and Foster, 1985).'

In Sapt Kosi river basin specially, riffles are wide and shallow, while pools are narrow and relatively deep. In Sapt Kosi river basin the lower catchment area where the channel bed is wider and sinuous course having pools. Similarly, channel depth is greater in at riffles in straight channel than in curved channel, where it is relatively shallower at pools in straight channel than in curve channel. Pool-riffle sequences introduce variations in channel forms. Channel width-depth and velocity all varies generally, it has been observed in Sapt Kosi river basin in its tributaries that pools have been developed at concave bank and riffles at coarser channel (Fig. no. 6.5) in sequential manner. During the Monsoon period when the channel of Sapt Kosi forming discharge (bankful) is associated with high and low flow velocities when in high velocity currents erode the pools while low velocity currents deposit sediments on riffles, this resultant into increased cross section of pools due to erosion by rapidly flowing water, ultimately reduces flow velocity while accretation of sediment on riffles results decrease in cross section and increase in flow velocity.

As per E.A. Keller (1972) five stages of model for development of pool-riffles sequence on the basis of evidences of regularity in the spacing of pools and riffles have been found as follows:

i) Straight channels normally having no pools and riffles asymmetrical shoals are associated with the deposition of sands.

ii) The small size pools and riffles associated with the width of the channel.

iii) Most of the riffles and pools are found in the lower catchment area of the basin and its sinuous course. Several point bars have been developed here.

iv) Pools and riffles are also found fully developed at the higher spacing areas.

v) Incipient pools and riffles are formed at the crossovers. It is thus, the channel bed topography of Sapt Kosi river basin is very complex. The lower catchment area demarcated by depositional landforms where as the upper catchment area erosional.

vi) The pools like feature have been developed on the channel bed be to interaction of channel flow and transport of sediments.

CHANNEL PATTERN: 'The channel pattern or map view of a river is usually considered as straight, meandering or braided (M. Morisawa, 1985). Channel patterns on several criteria viz. sinuosity, channel stability, number of channels, shape-discharge relationship, sediment-load type have been studied by several geomorphologists, as L.B. Leopold and M.G. Wolman, 1957; S.A. Schumm, 1958, 1972; A.D. Miall, 1978; B.R. Rust, 1978; D.I. Brotherton, 1979, R.I. Farguson and A. Werrritty, 1983.

Here we have attempted to apply the method of Leopold and Wolman, S. A. Schumm and

Leopold and Werritty has basically divided river channel of three types on the basis of sinuosity index. i) Straight channel (sinuosity index less than 1.0), b) Sinuous channel (sinuosity index between 1.05-1.5) and iii) Meandering channel (sinuosity index more than 1.5)

The sinuosity index of Sapt Kosi river basin has been found

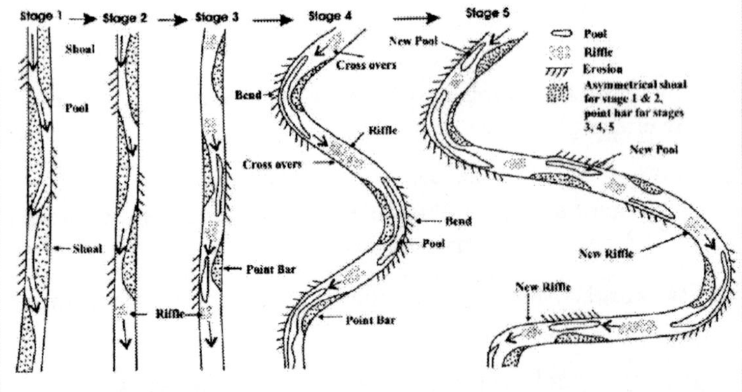

SEQUENCE OF POOLS AND RIFFLES IN SAPT KOSI RIVER BASIN

AFTER, R. I. Ferguson and A. Werritty, 1983.

Fig. 6.5

to be 1.02 and 1.15. It is less than 1.05 in the mountainous region and more than 1.1 in the plain region which indicates meandering or braided channel in the plain region while to some extend straight in the mountainous region.

As per S.A. Schumm (1963 and 1972) classification scheme Sapt Kosi river basin in the lower catchment area it is mostly stable, where W/D ratio is less than 2.1. Suspended load channels are found to be depositional. The suspended load are found more than 85%, bed load is between 0-15. On the basis of the above models of the classification the channel pattern of Sapt Kosi river basin may be divided into five types. Sapt Kosi river basin is also characterised by meandering channel as the pool-riffle sequences introduced variation in channel forms as meanders are often associated with a frequency of one riffle-pool per bend. Likewise, it is also an example of braided channel. In the lower catchment area of Sapt Kosi river basin the channel is characterised by bars and islands. A braided channel characterised by- a) Incompetence of the river to the transport bed load, b) Highly erodible banks, c) Fluctuation in discharge change, d) Steep channel gradient and e) Abundant and coarse bed load.

CHAPTER – 07

CHANNEL DYNAMICS

Channel migration is believed to be dependent on the variation in discharge from season to season. High variation is usually associated with significant shift in the deepest channel while low variation is associated with comparatively stable channel. 'Major westward diversion do not correlate with major earthquakes in the region, but seem to record a persistent tendency of the river when it abandons its channel to migrate to the previously unoccupied fan surface rather than shift eastward to a channel recently abandoned. Channel shifting occurs during monsoonal floods and commonly results in major human disasters (Nayak, 1996).' Repeated survey in the past 228 years record a continuous westward shifting of the Kosi river for 113 km across its fan, in a series of episodic diversions of up to 19 km/year (Wells and Dorr, 1987). 'Gohain and Prakash (1990) recognised remnant patches of Old Alluvial Plain (OAP) in the western part of the megafan. They reported that the rest of the megafan surfacewas covered by different geomorphic elements of the Young Alluvial Plain (YAP) and are characterised by grey unconsolidated sediments. The occurrence of a patch of OAP was inferred by them to indicate that the Kosi River, during its latest sweep, in addition to slow continuous shifting at times moved in discrete steps leaving behind remnants of earlier deposited sediments (Gohain & Prakash, 1990).' There may be some exceptions to this general observation. Relation may however be established between the ratio of maximum discharge to bank full discharge on one hand and lateral channel shifting on the other. In some cases, there is a very good relation between these two parameters. There should be clear distinction between temporary (chute cut-off) and permanent cut-off by using temporal satellite data. The above morphological parameters are studied to explain the unique characteristics and behaviour of the river and their bearing on various problems. The channel parameters, which need to be corrected / modified to find the solution to the 14 problems, may be identified and the extent, to which such corrections / modifications are required, may be spelt out.

The facts to be considered are those of the Himalayan valley plan and the sculpture, structure and rock types to the present; the asymmetrical area of emergence from the gorge outlet to the gradual definition of the remarkably symmetrical cone; and the directions of deposition and of drainage, ultimately to the Ganga near its southward bend into the Bengal plain. The types of sand carried down the gorge to the cone will be noted and the known history of the main courses outlined from before 1730 to 1940-145, and the works for comprehensive control, achieved by 1955-57 and projected.

The three main tributaries of the short reach of the river in the gorge, the Sevenfold or Sapt Kosi, are i) the Arun Kosi rushing straight down from the Tibet Plateau south past the foot of Everest, ii) the Sone Kosi flowing from the west and collecting the waters of four or five Great Himalayan valleys and then joining the Arun near the upper end of the gorge and iii) from the east the Tamur which drains the foot of Kanchenjunga and joins the Kosi just below the Arun-Kosi confluence.

The annual rainfall increases from 125 cm in the plain to over 175 cm in the Lesser Himalayas and more on the Lower Great Himalayan slopes, then decreasing both on the snowy mountains and intervening valleys and gorges, to a low mean in Tibet.

From the west flows the Son Kosi,.whose longitudinal valley is 200 km in length and some 110 km in breadth from the Mahabharat crest to the Great Himalayan water parting : 21,600 sq. km. from the east the Tamur actchment measures roughly 110 km each way: 13,200 sq. km. and from the north the Arun gather a Great. Himalayan run-off from a triangular catchment which is rather in the lee of the Lesser Himalayas, and from its long Tibetan catchment which is rather in the lee of the Lesser Himalayas, and from its long Tibetan catchment: a total of perhaps 26,000 sq. km. the as most of this has only half the Son Kosi valley's rainfall or less, one may guess at ratio of discharges corresponding to area of 21,670, 10,000 and 13,000 or 10,000 sq. km. the evidence from the long profiles of the Arun-Kosi suggests continuing uplift of the Himalayas (Wager, 1937) the Sapt Kosi would appear to have maintained its over-flows across the rising ridges of the Middle and Outer Himalayas (the Mahabharat and Siwalik ranges) at a low northsouth point or line of weakness and probable instability, marked to east mainly by a continuation of the metamorphic Mahabharat range, but to west by an outer range of friable Tertiary rock, of which an outlier, 30 km in length, overlooks the area of the Kosi's emergence.

Instability in the area was sharply brought to notice by the Bihar-Nepal earthquake of 1934, revealing stresses and strains from the Plateau Rim across the plain into the Himalayas. The epicenter lay on the plain near Himalayas, south of Everest (Brett, 1935). An earthquake may be thought of as a catastrophic release of stresses and strains of prolonged diastrophic nature. Although the Nepal Himalayas have not been fully surveyed the observations of an indefatigable swiss geologist, Hagen, are revealing and so also are surveys made by oil companies (Hagen, 1963).

The successive overthrust of vast Himalayn rock sheets, no doubt linked to massive under-thrusts of the Indian block below, produced breakage, fragmentation and landslide, with the rolling and rounding by rushing water into boulder and pebble belts, as renewed every year by running water along the Himalayas foot. This can be watched from the plunging boulders and widespread shingles of the Himalayan rivers to the steep cones of boulders and pebbles at the foot of any Siwalik torrent, both of them widening out into beds of sand, whether great or small. As overthrust and countering underthrust proceeded, the gently successively upliftedand simultaneously partially eroded-strata of the Ridges of the Siwalik ranges their erosion in turn strewed fresh pebble bends until the present Bhabar pebble belt.

With the rise of the successive pebble bed, stream flow was increasingly interrupted and fewer points of overflow were left.

Each overflow tended to cut down a gorge, but then power diminished and ceased along considerable lengths of the Siwaliks, leaving ever fewer Siwalik gorges. Between these have been found low saddles or cols. Former gorges are still have many river and stream gaps, the Mahabharat Range has far fewer and of course the Great Himalayas have fewest of all.

In a range such as the Himalayas we see that the main drainage from crestline to mountainfoot is considerably interrupted by the rise of the successive hillfoot belts as outer ranges. And in fact there is seen a river pattern of many plainward, Great Himalayan streams turned at right angles, or almost, by low outer or outermost range and ridge line. Nevertheless the main Himalayan slope not only asserts its gravity force where a longitudinal valley 150 km or more in length at last crosses the Outer Ranges onto the plain by striking gorge; it also modifies appreciably the strict parallelism of a longitudinal valley by edging its growing river against the Outer Range, eroding its northern foot increasingly until the gorge is reached.

The Son-Kosi, after turning southwards, is again turned due east for 45 km, hemmed in by a ridge of Siwalik rock, before reaching the present gorge. Southwards of this 40 km reach their rise long ridges of Siwalik rock one of which hems in the valley drained by the 56 km long Trijuga which joins the Sapt-Kosi below the gorge. On the east of this emergent reach of the Kosi no continuing ridge is found. But one would expect signs of continuing and recurrent uplift and in fact 16 km east of the gorge outlet (Dear Dharan) the alluvium is sharply uplifted and dips to the south and west i.e. to the Sapt-Kosi. Some 20 km further east (near Karwa) is an unmistakable overthrust of more than 20 meter of light sandstone of the central Siwaliks over the recent Upper Ganga alluvium (Hagen, 1963). This Siwalik overthurst and the uplift of the alluvium belong to the present time, and its catastrophic consequences are expressed in the westward shifting of the Sapt-Kosi. When the first comprehensive report on the "rapid westerly movement which can now be added to his findings (Inglis, 1944).

Although neither author points this out, one would expect Sapt Kosi on emerging from gorge outlet to begin laying down a cone such as would show fairly symmetrical concentric contours. Instead, as Inglis's may first showed, the slope of the gorge from the north is modified by a slope from the east, now throwing the river westward hard against the toe or spur of the Siwalik range and undercutting it. Where a rocky river bank resists scour, the river tends to cling to the bank, for there the current is forced into a curving, diving and deeply excavating flow. Whether or no the cause of movement westward in the first 35 km below the gorge is due to overthrust from the east, the river's demolition of the lower part of an ancient

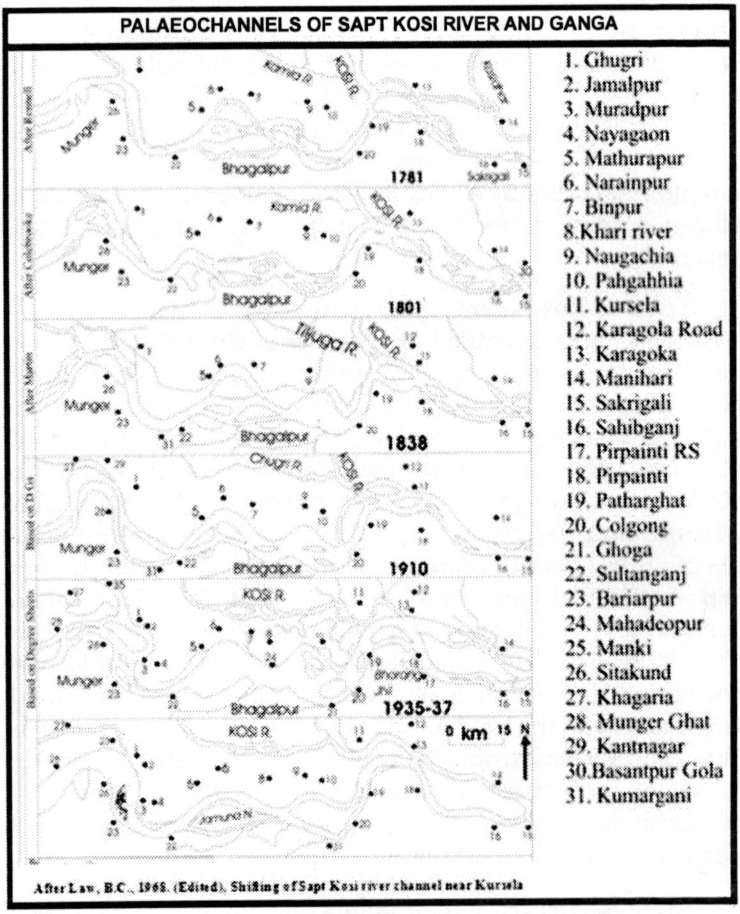

Fig. 7.1

embankment, the Bir Bund, which according to tradition in Bihar 'was built by giants', shows this westward movement to have occurred since historic or proto-historic times. The Bir Bund extended for 55 km with the dual purpose of containing irrigation in Nepal from a 50 km long eastward river and also of preventing possible westward movement in the upper Kosi in Bihar territory. The Bund runs due south for 30 km then slightly eastward and south again, this latter stretch controlling or guiding distributaries from the older, axial course. The contemporary breaching of this ancient dyke shows that for many centuries the main Sapt Kosi had flowed, like all rivers of the Indus-Ganga plain, fairly directly towards the sea, i.e. in the Punjab Sind, and from the Yamun's course to the Kosi towards the Bay of Bengal.

The cone of Sapt Kosi expands markedly from about 24 km below the gorge outlet and from this apex to the foot measures about 105 km. the 87th meridian is the approximate axis; the arc lies between 26°30' and 25°30'. The regularity and symmetry of

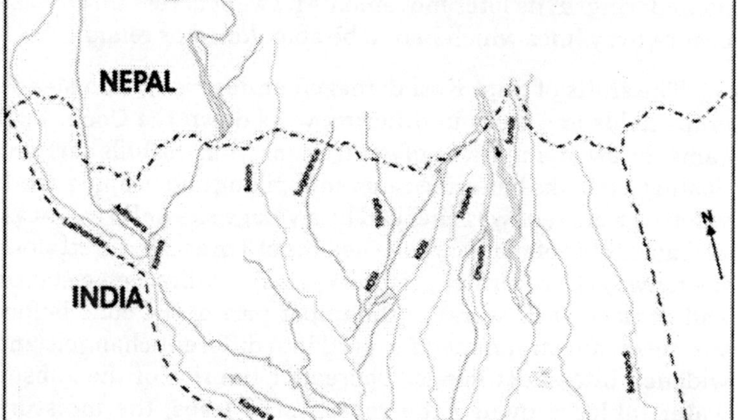

Fig. 7.2

the cone from its apex has been noticed above, although the Eastern Lateral collector flowing towards Bengal and the sea is naturally longer than the Western Lateral Collector. The shape and contours of the Sapt Kosi cone and its radial courses show that for millennia its deposits have been dropped successively from side to side. In the last two centuries the main channels have moved from east to west. This movement has been measured towards the foot of the cone, westwards along the latitude of Purnea, for 105 km (plus the distance to the Eastern Lateral Collector 27 km), movement from one lateral collector to the other totaled over 130 km.

From some times before the first map (1736) the distances westward in kilometers per year increased from about 1.5 km to the axial course in 1850 and thereafter at increasing speeds up to 1933-36 it moved 19 km. 6.5 km per year but after this the mean movement slowed down to an average of 1.5 km a year, when it reached the Western Lateral Collector in 1938. The river shifted in a series of swoops, covering a distance of some 19 km in a period of three years. Before control in 1955-57 the river course due south was 240 km compared to some 160 km fifty years before, owing to increasing deposition and consequent meandering. In its later movements it swept across three west-east railway lines which had to be abandoned or relaid.

The shifts of Sapt Kosi damaged or destroyed homsteads, crops fields and trees in differing ways down the Cone. First came the sweeping water of great floods from the hills carrying floating trees like battering rams and spilling and heaping sands where the current was checked by a village site or by west-east embankments of road or rail to heights of 3 m as measured along the railways from former ground levels. Such direct devastation and damage were worst in the upper part of the cone before the new main stream had fissured into different channels and widened into sheet floods. Thereafter the rise of the subsoil water tables drowned the roots of the trees, the moisture percolationg through the sandy soil as far as 8 or 10 km from its course. Where the subsoil water rose to within 0.6 m of the surface bulrushes appeared: where its was lower there grew tail grasses. At some distance from the river the water-levels in wells lay 6 m down but were higher nearer the river.

Table - 7.1
CHANNEL SAND MINERALOGY IN MEGA FAN AREA

River	Light Constituents	Carbonate Content	Heavy Constituents
Baghmati	Q:30-35% F:30-40% (KF>Plag.) M:5-8% (Muscovite) RF: 25-30% (Metamorphic)	2-3%	Major: Chlorite, Biotite (Bio>Chl) Minor: Tourmaline, Rutile, Actinoline, Iron oxide
Kamla-Balan	Q:30-35% F:30-40% F:30-40% (KF>Plag.) M: 10-15% (Muscovative) RF: 25- 30% (Metamorphic)	0-1%	Major: Chlorite, Biotite (Bio>Chl) Minor: Tourmaline, Rutile Actinolite, Iron oxide

Source: After Rajiv Sinha, 1995. (Q-Quartz, F-Feldspars, M-Micas, RF-Rock fragments)

In his 'Tent life in Tigerland' a New Zealander (I. Inglis, 1892) described the river as he knew it ." the main stream runs with a swift milky flood dividing the two great indigo and rice district of northern Bhagalpur and of Purnea. When swollen by the melting of the snows or by the annual rains, the river overflows its banks, and at such times the appearance of a broad swiftly flowing sea, for its breadth from bank to bank is often ten and in some places nearly twenty miles across. In the dry season, the watersalways of the same milky hue are confined to innumerable channels, some so shallow that the stilted plover can wade across, and others running deep and strong with a ceaseless gurgling swish that would sweep the stateliest elephant off its feet and carry its ponderous bulk far downstream. These streams seem to run at random over this deltaic plain, diverging here, reuniting there, forming a wide bend in one place, and cutting direct through the sandy soil in another. The face of the country is split up everywhere by a network of dry channels and shifting sand-banks; and over all, wherever there is an inch of soil the stately elephant grass spreads its feathery mantle".

As mapped in the successive folds of 1936, 1938, 1939 and 1949, the Sapt Kosi water spread out from the apex into wide ribbons of flooded land, perhaps twenty kilometers in width fissuring half-way down into two or more narrower bands and spilling into the widely flooded but calmer expanse of a spillhollow. The damage could spread beyond the high water by drowning of the tree roots, but when the floods were over the ribbon of irretrievably ruined land on the cone might be 110 km but from one to two kilometers wide at the foot. The final stages of slow, silting flood deposited a veneer of finer soil adaptable by reclamation to the making of new paddies.

A major factor in the violent movements of Sapt Kosi is the heavy load of sand it carries out from the gorge together with the size and the shape of its constituents. While the sand of the Arun averages only 0.14 mm and that of the Tamur 0.23 mm, that of the greatest of the three tributaries, the Son-Kosi measures 0.3 mm. the sands of Sapt Kosi described as 'white' by Inglis are in fact composed of grains of differing shape and colour, as yet unrounded by prolonged rolling and remaining angular in outline. While some fragments are of irregular form and are white, others are in thin flakes and are transparent or black. The last two being very shiny, catch the light: they are unmistakably mica somewhat as bits of paper are readily blown by the wind because their surface area is large and their thickness exceedingly small, so the flaky micaceous sands of the Kosi are uplifted and swept forward. On other cones, grains of equal or greater weight being of compact shape are less readily uplifted and they tend to interlock and to consolidate and in time chemical consolidation may follow. These Outer Hills disintegrate rapidly and the natural process has latterly been accelerated by temporary, sporadic clearings for cultivation which lead to rapid denudation and landslides so increasing the Son-Kosi's load. Inglis was the first in India to try to measure the relationship of three factors; discharge or flow of water, charge or load of sands and silts, and slope or gradient. It was his realization of the importance of Sapt Kosi's load which led him to try to measure the load factor. The much greater discharge of the Son-Kosi relative to the combined Arun and Tamur largely explains the predominance of the Son-Kosi's sands unloaded over the cone and its unstable river regime. An additional reason is the flakiness of its mica particles.

It should still be possible to recognize the sequence from scouring to silting off-take on the older cones. Grasp of the dynamic process will thus enable the comprehension and mapping of alluvial forms and associated textures of soils.

A that point Sapt Kosi had driven through the Bir Bund. Such an embankment, if built on land sloping towards it will prevent a change of course so long as it is maintained and its lower end is strongly held. As the embankment was built to protect the irrigable land watered by a minor tributary from the west, the Trijuga, its builders must have had reason to place it there along a length of some 55 km. but since for considerably more than two centuries it was not severely attacked, any repairs would have been locally made not organized by a strong government, whether in Nepal or Bihar.

As an unpublished 10ft contour map shows, for the 30 km. stretch from 350 ft. (105 m) near the gorge outlet to 250 ft. (75 m), where the convex curvature of the cone is about to begin, the slope of the area of emergence is now rather towards the head of the bund, so throwing the river against it and breaching it. This movement of distributaries from side to side of a cone has been likened to the movement of a pendulum swinging from a fixed point at the conal apex, yet where Sapt Kosi's movement over the last two centuries is concerned the comparison is misleading. The apex of the cone proper has been shifting

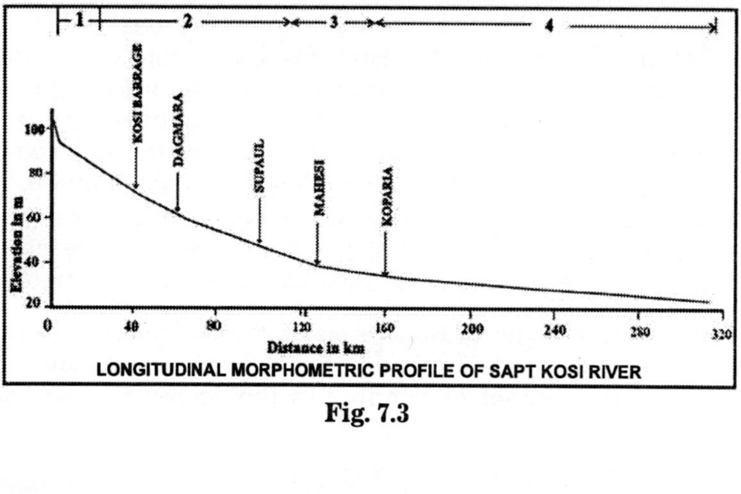

LONGITUDINAL MORPHOMETRIC PROFILE OF SAPT KOSI RIVER

Fig. 7.3

westward. The main flood channels have recently moved from east to west, from a leftward to a rightward course. For 24 km. below the gorge-outlet the main channel used to flow due south before, bending somewhat eastward of 100 km. to just west of Purnea town, recorded by 1736. The east most Lateral Collector, the Old or Burrhi Kosi, lies still further east. The main channel moved westward in successive phases to 1873-92. (The 1931 Census noted the westward migration of people from Purnea into Bhagalpur and within Purnea, the increase in the western part of the District. This was because of movement from the new Kosi to the old).

OXBOW LAKES: 'In many lowland river systems narrow channels connect the main stem river to lakes on the floodplain. Depending on the relative water elevations of the river and lakes, flow through these channels may be into or away from the river (Rowland, Lepper, William, Wilson and Sheldon, 2005).' Oxbow lakes are significant landscape founding in the Sapt Kosi lower mega fan area. The significant features are low laying surface with U shape, extremely slow flowing river water and meandering channels. River change their course almost in every year's maximum fluvial season. There are lots of evidences of cut off channels of Sapt kosi river creating a free-standing body of water through one can easily correlate the history of river's old path. Longest oxbow lake is Bakhtiarpur Oxbow Lake ($25^0$43'55" N & $86^0$33'08" E) having 8603 meter length and 263 meter width but this oxbow lake is in broken shapes. Naugachia oxbow lake has well defined U shaped oxbow lake (7226 m length). Similarly Kursela oxbow lake has well defined U shaped (5304 m), both oxbow lakes have developed huge settlement pressure on both side of the lake and these lakes are becoming as waste drainage system. These oxbow lakes are basically using as water irrigation in dry season, but in rainy season few oxbow lake converts into regular stream. Thus, in rainy season these oxbow lakes possess as dynamic nature as changing their shape and size regularly. Without the above mentioned there are more than 500 small oxbow lakes in Sapt kosi river mega fan area.

'Kosi river in north Bihar is one of the problematic rivers, which is responsible for frequent flooding in the region. To prevent the impact of flooding in the region number of

Table - 7.2

OX BOW LAKES OF DIFFERENT LOCATIONS IN KOSI RIVER BASIN

S. N.	Name of Oxbow Lake	Location	Elevation	Length	Width	Sapt Kosi Left/Right	Geometric Shape/ Pattern	Remarks
1.	Kursela oxbow lake	25°26'66"N & 87°15'03"E	113	5304	85	Left		settlement construction on left side
2.	Dhobinia milik dakhinwari	25°28'08"N & 87°13'35"E	106	3040	90	Left		Developed in broken shape
3.	Naugachia oxbow lake	25°23'13"N & 87°05'41"E	121	7226	88	Right		settlement construction on construment construction on the curve (both side)
4.	Partapnagar oxbow lake	25°23'13"N & 87°04'57"E	113	3385	145	Left		Agricultural field
5.	Baluachak oxbow lake	25°27'21"N & 87°02'17"E	110	4988	138	Left		Agricultural field
6.	Bishnupur milik oxbow lake	25°26'01"N & 86°58'24"E	108	1858	93	Right		Agricultural field
7.	Sheikhpura milik oxbow lake	25°34'60"N & 87°14'26"E	121	2706	50	Left		Agricultural field
8.	Bakhtiarpur oxbow lake	25°43'55"N & 86°33'08"E	133	8603	263	Left		Agricultural field with broken lake
9.	Naltigarh oxbow lake	25°49'31"N & 86°33'08"E	141	3385	171	Left		Few water mass & agricultural
10.	Gamrahu oxbow lake	25°48'47"N & 86°27'36"E	142	3969	91	Left		Extremely broken
11.	Matsyagandha oxbow lake	25°55'02"N &	146	6576	190	Left		Almost dry bed

		86°34'05"E						
12. Tarawe oxbow lake	26°01'43"N & 86°41'58"E	158	2114	174	Left		Almost dry bed
13. Jhagarua oxbow lake	26°01'10"N & 86°13'19"E	154	2255	110	Right		Agricultural field
14. Mansara oxbow lake	26°00'18"N & 86°19'33"E	150	1301	95	Right		Agricultural field
15. Dighni oxbow lake	25°35'07"N & 86°23'22"	126	6024	122	Right		Agricultural field
16. Kabilashi oxbow lake	25°48'40"N & 86°13'19"E	139	1731	142	Right		Agricultural field & settlement
17. Belahi oxbow lake	25°51'26"N & 86°11'40"E	146	1272	2004	Right		Agricultural field & settlement
18. Belgawun oxow lake	25°55'47"N & 86°09'48"E	158	1594	73	Right		Agricultural field
19. Lalpur oxbow lake	25°55'47"N & 86°11'18"E	152	2001	40	Right		Agricultural field in broken
20. Parsabirbal oxbow lake	26°14'60"N & 86°59'11"E	208	709	20	Left		Agricultural field in broken
21. Supali oxbow lake	25°36'16"N & 87°08'13"E	119	3017	209	Left		Almost dry bed & agriculture
22. Kalyanpatti oxbow lake	25°46'58"N & 87°08'13"E	154	1138	43	Left		Settlement & Agriculture
23. Fatehpur oxbow lake	25°47'17"N & 86°56'34"E	144	1138	43	Left		Agriculture
24. Mora oxbow lake	25°27'05"N & 86°44'58"E	113	2540	153	Right		Agriculture
25. Koila oxbow lake	25°27'05"N & 86°46'19"E	118	985	137	Right		Settlement beside left side

Source: Based on the topo map (72/K), Google Earth and Self Excursion in the river basin (Naugachia region of Bihar, India) embankment was constructed since post independence period, but the breach event occurred in 2008 in the eastern embankment, forced the hydrologist, geologist and engineers to think again in a different manner about the problem. After the breach event most of the Kosi water shifted towards eastward direction and it re-occupied its available paleochannel in the region to flow the water. Though after the breach event, once again the embankment was repaired and the main channel of Kosi was forcefully allowed to flow within the embankment. This solution is may be a temporary solution but it is not sure that this type of events will not repeat in the future (Panda, 2012).'

It preserves within its lithology and structures, different phases of geomorphological history and influence of different geomorphic processes. Sapt Kosi river basin is the most destructive river in India. Its location is mainly on the north Bihar plain creating long history of flooding. It causes death of indigenous people and unending immigration of native villagers from their home. It is destroying productive power of fields by depositing bunch of sediments, mostly sand through overland flooding and inundation. Mostly of the palaeochannels of Sapt Kosi river basin get rejuvenated during monsoon season and their flow in to the present channel of Sapt Kosi is also obstructed due to the embankment along the left bank. Infect, embankment in the left bank is too close to the active channel and the river has to plays to spread of its overbank flow. Sapt Kosi mega fan is very notable in the world for its geomorphic features.

Sapt Kosi catchment is fed by monsoon rainfall as well as snow melt. 10% of the catchment up to the Chatra is above perpetual snow zone of Himalaya. Sapt Kosi catchment gets rainfall in the monsoon season and accounts for 80% of annual rainfall. 74% of the discharge of the Kosi can be accounted for by the precipitation in the form of rainfall. The average annual runoff of Barashetra is 553043 mm3 out of which 80% is contributed from June to October. The minimum annual runoff at the same place is approximately 38.83 mm^3. Sapt Kosi is a consequent stream; its tributaries are also consequent. But the master stream of Sapt Kosi is to some extent subsequent in the upper reaches. The general pattern of drainage is dendritic but some other patterns are also absorbable such as radial, parallel, directional trellis, braided, anastomatic, complex trellis, centripetal etc.

Analysis of peak flow in Sapt Kosi indicates that the peak flow can be 10 times as large as the mean discharge in a single year. There are 8 sediment and gorge discharge observation sites. The maximum flow observed at Barashetra was 25,880 m^3/s in 1998 and water surface elevation for this discharge observe to be 132.18 m^3/s. The analysis of sediment load carried on Sapt Kosi for the period 1948-81 has revealed that on the average it carried 95 Mm^3 of sediment annually, of which course, medium and fine size materials are 18.95, 25.11 and 55.94% respectively. Analysis of sediment size and slope of different sections has been observed and reveals the fact that the longitudinal sections of Kosi river as obtained in 1995 between Chatra and barrage at Bhimanagar the average bed slope is 9.38×10^{-4}, whole downstream for the 64 km the average bed slope is 2.70×10^{-4}.

Sapt Kosi river basin in the lower catchment area forms wider beds of meander which is indicative of the dominance of geohydrodynamic factor and irregularity carried by Redelom topographical, sedimentological or artificial disturbances. The length of overland flow of entire Sapt Kosi river basin quite varies. The upper catchment area of the basin area is highly vulnerable in comparison to lower catchment area. It is observe that the stream junction angle of the different orders of the basin varies the stream junction of the 4th order is approximately up to 90^0.

As intensive study of the hydrological studies of the basin reveals that the nature of the fluvio-environmental condition and hydrology of the basin is very complex. The changes in the region have generated a number of hazards like mass erosion, rill and gullies erosion, water logging and floods. Owing to high degree of slope the current in the river basin signifies by high flow-discharge and erosion. Cutting of meander slopes for straightening the course of the river channel. Channelization, diversion of the discharge of the channel in the raining season and grading of river bed slope. Construction of symphons, remodeling and reconstruction of road-bridge over the channel, from source to mouth enhancing the discharge of the river. The valley side slopes along the river are convexo-concave. Convexity in slope is a temporary state of slope development at the end of a spur by rapid floe of water current. The slope more than 40⁰ marked by scarp. The slope having 35⁰ to 40⁰ are having steep freefaces. These slopes are mostly found in middle and upper Himalaya. The slopes having angle of 30⁰ to 35⁰ are constant or repose slopes, which have been formed the accumulation of debris and they are more common on the accumulations of colluvial fans. The slopes with an angle of 20⁰ to 25⁰ are mostly agradational slopes. The slope having angle of 15⁰ to 20⁰ which are mostly formed in the Terai region are considered as moderate aggradational slopes where as the slopes below 15⁰ are terrace slope. The slopes between 5⁰ to 15⁰ are depositional slopes.

The suspended sediment record suggests considerable spatio-tempo variation in the concentration of load and yield. During the monsoon season at all the sides the coefficient variability is more than 20% indicating significance fluctuation in the concentration in the level. The sediment yield, as expected decreasing with increase in catchment area and suggests storage at sediments within the basin. There is a moderate to good relationship between discharge and suspended sediment concentration. It is indicative of more suspended load is transported by high flows and than by low flows. Its depend upon sediment supply along with the magnitude of discharge. Large floods with a recurrence interval of more than 6.93 years carry between 20% to 70% annual suspended load.

There is close relationship between elevation and discharge. The higher catchment area of the basin possess low amount of discharge where as lower elevation of the catchment area possess higher values of discharge. Instead, data belonging to various hydrological stations situated in the basin area such as Uwa Gaon (1,294 m), Beni (2,350 m), Helambu (2,134 m), Panauti (1,480 m), Budhanilkantha (1,454 m), Gauri Ghat (1,300), Shyamdado (1,660) and Sundarijal (1,600 m) having 192, 6.15, 7.14, 2.26, 165.8, 2.32, 0.41, 0.41 m^3/s of discharge respectively, which possesses very low amount of discharge due to high elevation. Where as in Chatra (183 m), Baltara (50 m), Kampughat (200 m), Mainachuli (125 m), Simle (151 m), Mangalpur (177 m) and Pandhera Dobhan (180 m) having 1,589, 2236, 966, 76.5, 444, 146.4, and 85.5 m^3/s of discharge respectively. The average discharge of the river is 2,06,347.4 m^3/s.

Mean monthly sedimentation at Sapt Kosi barrage near Hanumannagar is measured maximum in July month in 2008 it was more than 8,000 ppm. Due to sedimentation the changing behavior of the river course are occurred. Such type of behavior of the channel are seen near the confluence of Ganga and Sapt Kosi in Kursela when Google earth images are differ from today's situation (see Fig. no. 10.15; till 15/04/2014).

Channel dynamics referred as the processes of landform development rather than the forms they produce. In the Kosi river basin the main outline of the present relief are related with the micro physiographic categories. These outlines of the relief have been fashioned by the channels of ranging water. Almost all streams/river problems geomorphic hazards and resultant dangers may be avoided or minimize at the river corridor is considered as public domain and is protected as a vital resource. The channel of the Kosi river particular in plain in Bihar is highly dynamic as it has shifted the main flow 120 km during 250 year. The flow is steady and constant along the channel. The channel is wide and channel slope decreases according to the relationship. The bed material at any section of the reach has the same size and gradation as the bed material discharge in the upstream section.

CHAPTER – 8

FLUVIAL DYNAMICS

'The fluvial dynamics in a river basin refers an analysis of the characteristics which provides important information on erosion, transport and sedimentation processes. These processes are closely linked to the morphological characteristics of the bed and to the evolutionary dynamics of the water course (Billi, P. & Paris, E., 1992).' In particular, the fluvial dynamics of the Kosi river basin is very evolutionary and limited to a great extent, from a planemetric point of view. 'The dynamics have therefore been mainly of an altimetric type and consist of a generalized lowering process of the bottom (Mosely, M.P. & Tindale, D.S., 1985).' A generally high sedimentary deficit was found, evidence of this being the serially or absent of sedimentary bed erosion.

Sapt Kosi river basin has a mobile predominantly gravel bed in the Middle and the Shivalik Himalaya which, in its natural conditions, presented typologies which varied form braided to winding forms with considerable mobility and abandon bars. In the present study mobility of the bed is often totally impeded by channelization and planimetric and altimetric fixing works. Such as bank protections or weirs, which the flood hydrographs and sediment discharge are also considered by the hydraulic works.

FLUVIAL DYNAMICS: Fluvial dynamics is considered as one of the significant measures of geomorphic effectiveness. Present exercise is devoted to the analysis of river dynamics as well as the characteristics of the suspended sediment transportation for the monsoon dominated region of Sapt Kosi river basin. The principal characteristics of the suspended sediment concentration, suspended sediment and sediment yield during the monsoon season in Sapt Kosi river basin have been provided.

Time series plots for one year wise to where constructed to get an idea about temporal variations in discharge and sediment concentration. The heterogenty of Sapt Kosi river basin characterized by great complexity in its physical aspect as well as physical complexity. Here we have tried to present an effective analysis of the fluvial dynamics and sedimentation for the analysis of fluvial dynamics the following important parameters have been considered.

Water discharge in a river basin is one of the significant measures of geomorphic effectiveness. The present chapter is devoted to the analysis of discharge status of the some of selected rivers of Sapt Kosi river basin. It is also an important river mechanism which yields several geomorphic attributes and the process of landscape development. The volume of discharge of the different rivers of the basin has been analyzed to find out the discharge attributes of the basin.

River discharge is intimately related with the surface configuration and geologic attributes in a region. The entire Sapt Kosi river basin has been subject of study by Bordet, Gansser, Hagen and Akiba et. al.

DISCHARGE IN THE KOSI RIVER BASIN: Sapt Kosi catchment lies in the monsoon belt and the whole catchment are fed by rainfall and as well as snow belt. Sapt Kosi catchment gets rainfall due to monsoon which begins around June and retreats in the middle of October. This accounts for 80% of the annual rainfall. During April and May thunderstorm occurs in the catchment. The annual rainfall decreases 1,200 mm at the foothills to 350 mm on the southern slopes of the Himalaya. In Tibet catchment areas it is about 250 mm while about the lower part of the Kosi catchment it varies from 1,380 mm to 1,500 mm. July and August provides the maximum rainfall (Mukherjee and Aich, 1963) have estimated that 74% of discharge of Kosi can be accounted by the precipitation in the form of rainfall.

Analysis of peak flow in Kosi indicates that the peak flow can be ten times as larger the mean discharge in a single year. Flow duration curve for the course for the Kosi at Barakhetra given by Gohain and Prakash (1990). Its approximate flow

Table - 8.1

FLOW DURATION IN SAPT KOSI RIVER AT BARAKHETRA

Monthly average discharge in m^3/s	300	400	600	700	1400	3600	4300	4800	5800
Percentage of time equaled or exceeded	100	80	60	50	40	20	10	5	1

Source: CWC, Govt. of India, 2009

duration are given in Table no 8.1.

DISCHARGE IN THE DIFFERNET RIVERS OF SAPT KOSI BASIN: For many years, agencies of governments have maintained gauging stations along rivers all over the world. 'All these stations, usually on segments, or reaches, that are straight and have carefully measured and relatively stable cross sections, the discharge, water-surface level, channel shape, stream velocity, amount of dissolved and suspended mineral sediment, and other variables are periodically or continuously recorded (Dingman, S.L., 1994).'

Discharge (Q) is the amount of water passing through a channel cross section in unit time. It is defined as $Q = v\text{A}$

Where , v is the average flow velocity through a crosssection of area A. In channel flow, area is the product of channel width (w) and average depth (d), and the definition of discharge is commonly written as $\text{Q} = wdv$

Q is expressed in units of volume per unit time, typically, m^3/s. To determine Q with necessary precision, depth is measured at close intervals across a channel, and the velocity at several depths is measured at each interval. Then w,d, and v are calculated for the particular cross section.

'A graph of discharge plotted against time at a gauging

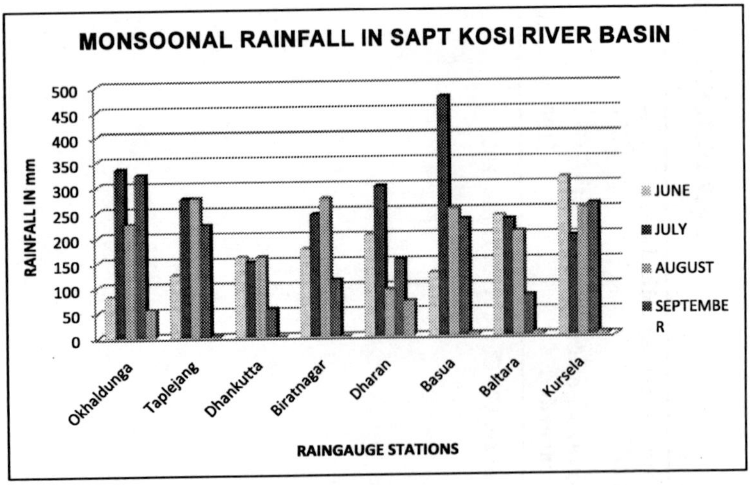

Fig. 8.1

station is called a hydrograph (Clozier, R., 1940).' Depending on vegetation, soil type, infiltration rate, and a host of other factors, the hydrograph reflects the precipitation in the watershed above the station. Stations progressively downstream have broader hydrographs with flatter crests at progressively later time, depending on the time required for the storm water to reach the station and various temporary storages within the basin.

There is close relationship between elevation and discharge (Fig. no. 8.2). The higher catchment area of the basin possess low amount of discharge where as lower elevation of the catchment area possess higher values of discharge. Instead, data belonging to various hydrological stations situated in the basin area such as Uwa Gaon (1,294 m), Beni (2,350 m), Helambu (2,134 m), Panauti (1,480 m), Budhanilkantha (1,454 m), Gauri Ghat (1,300 m), Shyamdado (1,660 m) and Sundarijal (1,600 m) having 192, 6.15, 7.14, 2.26, 165.8, 2.32, 0.41 and 0.41 m^3/s of discharge respectively, which possesses very low amount of discharge due to high elevation. Where as in Chatra (183 m), Baltara (50 m), Kampughat (200 m), Mainachuli (125 m), Simle (151 m), Mangalpur (177 m) and Pandhera Dobhan (180 m) having 1589, 2236, 966, 76.5, 444, 146.4, and 85.5 m^3/s of discharge respectively which is evident from the above graph's central tendency. About 49×109 m^3 of water flows into the Ganges River and merges into the Indian Ocean every year (Gleick, 2003). But in 2009 it was 1,560 m^3/s water average discharge in Sapt Kosi river near

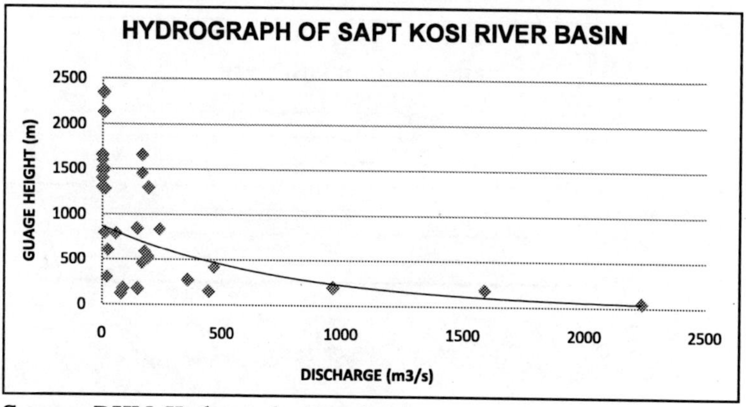

Source: DHM, Kathmandu, Nepal

Fig. 8.2

Chatra (data displayed on the notice board of Chatra hydrological station).

Table - 8.2

YEARLY MEAN DISCHARGE OF DIFFERENT TRIBUTARIES

S.N.	RIVER	YEARLY MEAN DISCHARGE (m^3/s)	LOCATION
1	SAPT KOSI	1,310	CHATRA-KOTHU
2	SAPT KOSI	2,236	BALTARA
3	ARUN RIVER	192	UWA GAON
4	ARUN RIVER	466	TUKEGHAT
5	ARUN RIVER	444	SIMLE
6	SUN KOSI	553	KHURKOT
7	SUN KOSI	176	PACHUWAR GHAT
8	SUN KOSI	716	HAMPUACHUWAR
9	BHOTE KOSI	239	BARABISE
10	TAMA KOSI	145	BUSTI
11	KHIMTI KHOLA	19	RASNALA VILLAGE
12	BALEPHI KHOLA	56	JALBIRE
13	BAGMATI	67	BHORLENI
14	BAGMATI	28	GAURI GHAT
15	BAGMATI	85	DOBHAN
16	KOSI BARRAGE	3,29,4827	HANUMANNAGAR
	MEAN	**2,06,347.4**	-

Source: DHM, Nepal & CWC, India, 2009

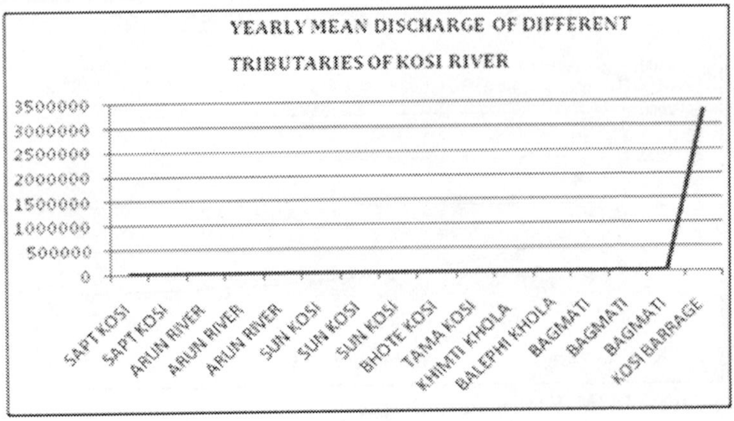

Fig. 8.3

It is evident from the above table (Table no. 8.2) that the mean discharge of Sapt Kosi river basin is 2,06,347.4 m³/s. Highest yearly mean river discharge is found in Kosi barrage near Hanuman nagar (Nepal), it is 32,94,827 m³/s. It is interesting to note that Baltara is located in Terai region that is why the hydrological station got its optimum water resource from the upper catchment area where as lowest water quantity is found in Khimti Khola near Rasnala village because it is located in the upper catchment area of about low precipitation region.

PEAK DISCHARGE IN SAPT KOSI RIVER BASIN

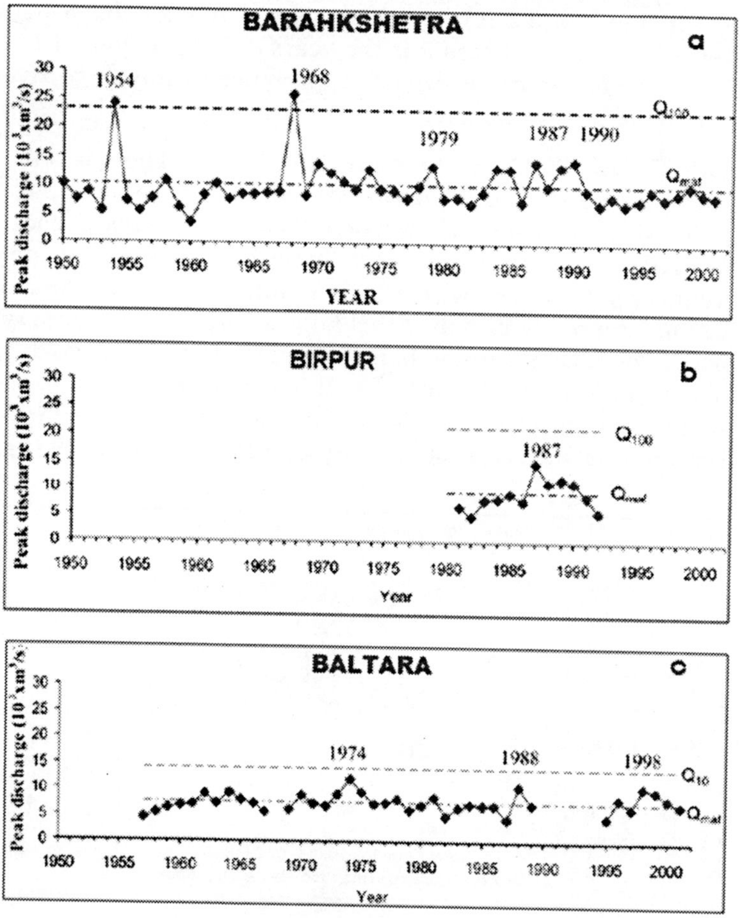

Fig. 8.4

Yearly mean discharge of different hydrological stations can be also seen in the below Fig. no. 8.3. In this graph presentation the highest values are of Terai region where as lowest amount of discharge is mostly found in mountainous region.

Peak discharge are recorded in 1954 and 1968 in Barahkeshetra hydrological station wehre as in the year of 1979, 1987 and 1990 are more than average yearly river discharge recorded. In the below Fig. no. 8.4 (b) it is the year of 1987 when peak dischage have beene recorded in Birpur hydrological station.

In Fig. no. 8.3 (c), the peak river dischage have been recorded in three times it is the years of 1974, 1988 and 1998 where as maximum years are experienced as average river discharge in Sapt Kosi river.

RIVER DISCHARGE AT KOSI BARRAGE: There is only a single big dam constructed over Sapt Kosi near Hanumannagar in Nepal with the help of Govt. of India during late five decades. Records available for some years in the barrage's hydrological department by whom regularly monitoring discharge and sedimentation. It is evident that huge amount of discharge are occurring which is more than 80,000 m³/s, few discharge daily also recorded up to 10,000 m³/s. It is again clear (Fig. no. 8.5) that maximum river discharge occurs during maximum fluvial season, it is July, August and September month of 2009 year,

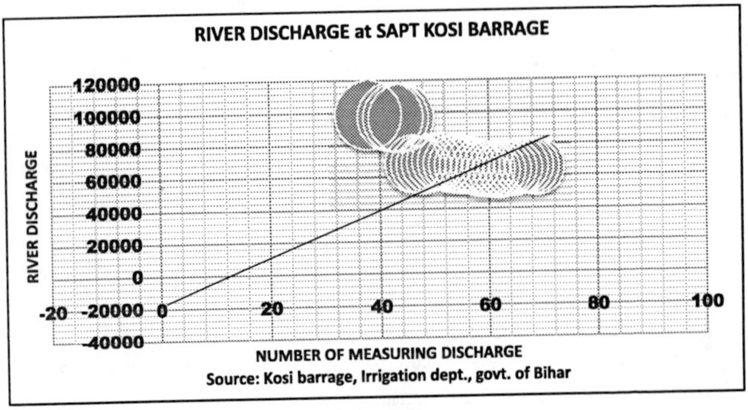

Fig. 8.5

whereas during winter season the river discharge is very low.

THE DEBRIS LOAD OF RIVER: Debris flows in Sapt Kosi river basin can be classified into four types. Tibetan Himalayas is extremely dry and its physical weathering is very intense. Small-scale rainfall triggered debris flows are common. In the high lands, debris flows are triggered by glacial lake outbursts, but they occur infrequently. In the middle of the basin, rainstorm debris flows are widely distributed. In this area, there are rich loose materials from metamorphite slate, schist and phyllite. These metamorphites offer a large amount of solid materials for the debris flow. In the meantime, summer rainstorms are highly concentrated from June to August (over 70% of the annual precipitation). In addition, the dry soils are highly collapsible and changeable under the action of rainfall, which could trigger the outbreak of geotechnical mudslides easily. There are more torrent debris flows in the Small Himalayas. The rainfall flood-debris flows, in Tibetan Himalayan, concentrate along the Sino-Nepal highway. Figure (gullies erosion. In sino-tibet road) illustrates the Menbudui gully debris flow field investigation (28°44'56.00" N, 86°09'38.30"E). The area of the gully is 2.3 km^2 and the elevation drop is 450m. The bedrock of the gully is 13m slate and metamorphose stratum. In the debris flow deposit, the gravels with more than 2 mm sizes are 62 %, the sands of 0.052 mm are 24 %, the powder grains of 0.010.05mm are 3 %, and the clay particles with less than 0.005 mm is 6 %. There are 1.25 million m^3 of loose solid materials in the investigated site and they occupy the terrain map of Menbudui gully along the Sino-Nepal highway. 5.3% of the whole area. The gully has 12 debris flows every summer. The debris flow alluvial fan covers an area of 875 m^2. The Zhangzangbo debris flows are a type of the Himalayan glacial lake outburst debris flows located at Nyalam County, on the southern slope of Himalayas. The Zhangzangbo ditch (28°40'46.20"N, 85°59'59.700" E, and its outlet altitude at 3,259 m) was a tributary of the River Boqu with an 8.5 km long channel and a drainage area of 50.5 km^2. The highest point of the catchment is 6,109 m and the outlet attitude to the Boqu River is 3,168 m with a relative relief of 2,941 m. The glaciers and permanent snow covered an area of 27.7 km^2 (45.15% of the catchment). The average slope of the hillside above the road is

> 40^0. At an altitude of 3,550 m above the valley, including seven major tributaries, the vertical channel gradient is large, about 149.6% and 478.9%. There are many high mountains, deep valleys and steep slopes in the Zhangzangbo catchment. The exposed strata are Presinian Kornhill Bridge Group (AnZk) with quartz mica schist, biotite gneiss, and gravel of Quaternary slope deposits from soil residual. Landslides and avalanches are widespread. The Zhangzangbo gully moraine platforms are widely distributed on both sides. The modern glacier tongue and front surface are covered by lateral moraine and end moraine. The arid valley of the middle Sapt Kosi River mainly suffers from rainstorm debris flows. A typical example is located at the Khurkot debris flow gully. The basin area of the gully is 3.8 km^2. There have been a large number of landslides in the middle-upstream of the gully. In the past, the debris flow activities had led the river to change its courses. The maximum debris flow boulder reached 4.8m×2.4m×1.9 m. The flood and debris flow lasted for 34 days. The debris flow contained a lot of stones, wood and roots. The debris flow speed is equivalent to people's running speed. According to the clay content analysis measured in the debris samples, the debris flow density was about 1.81.6g m^3s. The peak flow in the measurement section was 80100 m^3s and the velocity reached 56ms. The accumulated debris volume reached 30,000 m^3. In the small Himalayas, debris flows caused by torrents are quite common. They often threaten roads, farmlands, residential areas and other infrastructures and lead to the damage of forest along the way (N. Sh. Chen, 2013, Water hazards in the Kosi River basin, pp. 802-803). Many region of upper region of Tribeni experiences such type of debris flow especially in Tamur river basin where very loose material such as limestone and slate are available. Landslide and avalanches are common in this region which is basically the source of debris materials.

BED LOAD AND SUSPENDED LOAD: Sediments in the rivers are two types that are suspended sediment and bed load. Suspended sediments are fine and are mixed with water and their percentage is quite low. The suspended sediments in the rivers are added by glaciers, or by channel and side erosion and normal drop from the terrain either gully erosion or sheet erosion. River carries these sediments by their attractive force.

Most of the suspended load of silt 1.5% of the river with average silt content is 0.25%.

In maintains the general gradient is 200 feet per mile which means that the erosion power is tremendous due to steep gradient. Furthermore the gully being narrow it is further increased. But due to hardness of rocks, which are found in the mountainous region, it has become very difficult for the rivers to make heavy cutting in most of the places. In areas where phyllite and schist rocks are exposed, heavy cutting of wide valleys are found. In wide valley on the flood plain sedimentation is observed in the midland the rock being soft and due to deforestation by heavy population, the fine debris and coarse sands is added as land slip; slide as well s gully erosion. Fortunately the river in the E-W direction has less gradient i.e. 30 to 50 feet per mile and the sediment carrying capacity is limited, hence only suitable to that velocity of 3 to 4 meters per second is washed out. With the result by the time the river reaches Terai plain the sediments consist of sand and silt. In case of Churia river, rock being soft the detritus mostly sand and cobbles are found. All the rivers of eastern Terai brings heavy amount of sand in the flood and covers the entire area. The moment these rivers are low, they dump the sediment and in next flood they take different course and meandering begins. In the lower zone of Terai region most of channel have sand deposit and the bed has risen either few feet above or equal to adjoin land surface. The channels are rising near the foot hill due to sediment and the area covered by sand fan increasing. Terai region are characterized by 8-10 km of thick forest cover which let to save from the sedimentation carried out by coming rivers. These important forest cover can be seen in road from Dharan to Chatra, where each river channel is above the ground and the forest belt is working as a barrier.

The size and type of sediment are dependent on sources of supply and river character. In high velocity rivers the boulders are carried away up to 10 to 20 mile, whereas, in low it is carried out 5 to 8 miles. Quartzite and gneiss being hard they are rolled in the bed channel and carried out as far as 50 miles from their source. In Kampughat the boulders at Sunkosi bank are mostly quartzite and gneiss which appears in the northern part of

Himalayas.

COMPUTATION OF SEDIMENT LOAD: Sedimentation is the tendency for particles in suspension to settle out of the fluid in which they are entrained, and come to rest against a barrier. This is due to their motion through the fluid in response to the forces acting on them: these forces can be due to gravity, centrifugal acceleration or electromagnetism.

Table - 8.3
SEDIMENT LOAD

S.N.	River	Catchment Area	Mean Discharge	Annual Sediment Load	Average Longitudinal Gradient
		(km^2)	(m^3s^{-1})	($10^6m^3s^{-1}$)	(%)
01	Sun Kosi	18800	471	54.2	0.0062
02	Arun	34300	451	34.6	0.0134
03	Tamor	5800	347	29.6	0.0276

Source: N. Sh. Chen, 2013, Water hazards in the Kosi River basin, p. 797.

MEASUREMENT OF SEDIMENT LOAD: Sediments come in all shapes and sizes. Sediment sizes are classified by separating them into a number of groups, based on metric measurements, and naming them using common terms and size modifiers. The terms, in order of decreasing size, are boulder (> 256 mm), cobble (256-64 mm), pebble (64-2 mm), sand (2-1/16 mm), silt (1/16-1/256 mm), and clay (< 1/256 mm). The modifiers in decreasing size order are very coarse, coarse, medium, fine, and very fine. For example, sand is sediment that ranges in size from 2 millimeters to 1/16 mm. Very coarse sand ranges from 2 mm to 1 mm; coarse from 1 mm to 1/2 mm; medium from 1/2 mm to 1/4 mm; fine from 1/4 mm to 1/8 mm; and very fine from 1/8 mm to 1/16 mm. Unfortunately, the entire classification is not as consistent as the terminology for sandnot every group includes size modifiers. Sedimentation processes and quantity can be detail seen in Sedimentation chapter.

THE STREAM BED AND ITS DESCRIPTION: Once the critical shear stress on the bed in unidirectional flow is exceeded, the sediment particles forming the bed are transported at a rate, which increases with increase in shear stress on the bed. The bed in the process remains plane under some conditions, but under other conditions develops transverlsy oriented bed features known as ripples, sand waves or dunes, and antidunes as observed by Blasus, Cornish, and Gilbert and recently by Simons et. al. these bed-forms travel beneath the flow, take part in the sediment transport, and govern the relationship between flow velocity, flow depth and slope. In other words, they affect the friction and sediment transport. They also leave back a characteristic imprint in the enclosed deposits.

The stream bed of Sapt Kosi river basin is very significant to describe. Though it lies in three climate zone so the influence of those climate are clearly seen in the river bed. In the freeze region of high Himalayan mountain there is moraine distribution which is actually a sign of glacial work. In the Middle and Shivalik region of Himalaya their fluvial impact are clearly seen as boulder and pebble in the channel bed. Where as in the Terai region due to low gradient immediately big boulders last destination end up and after this in the mega fan area of Sapt Kosi big sand particles arranges in very sequence manner which is actually creates infertile soil. After that medium fine and fine sand particle creates middle and lower Sapt Kosi mega fan region respectively.

Fig. 8.6

The stream bed of Indrawaty river: The meeting place of Indrawaty and Bhote Kosi in Dolalghat, Nepal is significant to describe the fluvial work and their bed-form materials. Bed-forms of Indrawaty river near Dolalghat have been identified by the researcher. There is basically both channels of the river Indrawaty and Bhote Kosi in Dolalghat where a long and triangular boulder bar created by both stream. Usually the bed was covered with boulders in both marginal portion of 20 to 30 cm in size. In the middle of these the size of boulders becomes smaller and in the middle of the bar there was deposition of sand materials. In the end of the boulder bar the deposition of sand is also found.

SEDIMENT SIZE AND SLOPE: Garde et. al. (1990) have analysed the data collected from three boreholes at different cross sections of the now embanked Kosi. These bore holes were at the left, right and the center of leveed portion. The Central Water Commission has divided the entire reach of Sapt Kosi in four segments and the slope in each reach has been given as follows during 1982.

The longitudinal section of Sapt Kosi river as obtained and it can be seen that between Chatra and the barrage at Bhim Nagar, the average bed slope is 9.38x10-4, between the barrage and up to next 32 km downstream the average slope is 5.5x10-4, while downstream for the next 64 km the average bed slope is 2.70x10-4.

CHAPTER – 9

GEO-HYDROLOGICAL STATUS

'Basin hydrology is the scientific and quantitative explanation of the movement of inputs of water distribution and quality of water including the hydrologic cycle, water resources and environmental watershed sustainability (www.en.m.wikipedia.org/wiki/Hydrology, 2015).' hydrologic characteristics of a river basin played very significance role in shaping the hydrologic behavior as well as the geomorphic landscape.

In hydrologic scenes, however, prior to World War II, most basins were described in qualitative terms such as well-drained or poorly drained or they were connoted descriptively in the Davisian scheme as being youthful, mature or old. The mechanism of how river channels or networks actually form or how water gets into a channel was understood only in weg by hydrologists. Realizing this early 20th century view of streams and drainages it is startling to examine the Arant-Garde approach presented by R.E. Horton. His attempt to explain stream origins in mathematical term and to describe basin hydrology as a function of statically laws cited as the birth of quantitative geomorphology. 'However there can be little doubt that modern geomorphic analysis of drainage basins has its roots in Horton's origin work and his thinking was instrumental in the rise of new breed of geomorphology (Hortons, 1945).'

The ultimate source of river flow is, of course, precipitation which represents the major influx of water to any drainage basin. precisely, how much of that precipitation actually becomes the part of stream flow and what route of particular stream flows to reach a channel and how it is an instrumental in the development of geomorphic landscape are topics of great concern of hydrology. Infect, they constitute which is commonly referred to as the slope hydrological cycle. The present chapter has been dealt with the hydrological characteristics and the slope characteristics of Sapt Kosi river basin.

Sapt Kosi river basin is the most destructive river in India having a long history of flooding. It is not only a crush for the indigenous people but has destroyed productive power of field by depositing bunch of sediment mostly sand through over bank flooding and inundatation. Many of the paleochannels of Sapt Kosi river also get rejuvenated during the monsoon, so their flow into the present channel of Sapt Kosi river is also obstructed due to the embankment's along the left bank. Infect, the embankment along the left bank is too close to the active channel and river has no place to spread it's over bank flows.

The hydrological attributes of Sapt Kosi river basin is significant not only that the basin gets flooded and becomes devastative but also for the several features. All the hydrologic aspect like slope of paleochannels, flow direction identification and data generation is obtain from IGIS (version 1.1) and extracted from data IGIS about the cross section profiles, longitudinal profiles and slope value. Sapt Kosi river's main channel flooded water is most significant hydrological aspects which has attracted the hydrologists and geomorphologists from the beginning.

The inputs of water in any river basin are mainly two sources, i) Rainfall, and ii) Snowmelt water. Sapt Kosi river is fortunate one which receives water from both sources. The river becomes bankful during the monsoon period. The whole river regime flooded and receives high amount of rainfall water during this period. So, the hydrological parameter in Sapt Kosi river basin is mainly concerned with rainfall. The whole catchment area of Sapt Kosi river basin is 91,476 sq. km which receives the precipitation during the year flow in the river, size of channels, gradient and sediment load, topography and types of rocks are govern by the amount and intensity of rainfall received by the basin.

The amount of precipitation may be measured by a simple arithmetic method i.e. total basinal area by the amount of rainfall received during the single fluvial year.

Total basinal area: 91,476 km^2

Total rainfall: 1,930.3 mm

Water volume: 1.75657e+11 m^3

The other sources of water are snow, most of rainfall and snowfall occur in the basin in the month of June to September and enhance, discharge rises firstly in this period. In winter, only the bank storage, snow melting and ground water meets the need of the low flow. The total snow covered catchment area of Arun river is 4,475 sq. km.

Rainfall does not usually, strike the ground surface directly become its fall is impeded by leafs or trunks of the vegetal cover. The whole process is call interception. Losses are quite variable because they depend on numerous meteorological factors, vegetation types, landuse and seasonality. In addition, losses are dependent on storm duration, being high initially and decreasing at the storm continues. During long storm, the vegetation may eventually become saturated and all addition rainfall is passed to the surface. Rainwater that reach ground surface they still be presented from becoming part of stream flow because it is susceptible to further loss by evapotransporation.

INFILTRATION: Hydrologists have always being concerned about how much runoff will issue from a precipitation event and how quickly the runoff will enter into the stream channels. These factors determine if and when a flood will occurs and how high the river level will be at peak flow. 'In 1933 Horton suggested that the primary control of storm runoff is how easily rainwater can sink into the material on a sloping surface (A process known as infiltration) as precipitation falls on the slope surface, the water is absorbed in the ground at a rate called the infiltration capacity, which is a function of number of soil texture and structure, vegetation and the condition of the surface prior to precipitation (Hortons, 1933).'

Sapt Kosi river basin is highly heterogeneous in regard to the geological structure, slope, vegetation cover and soil texture etc. infiltration capacity also varies in different part of the basin. 'Infiltration in river channel in mountain is not a loss and as it returned into the same channel after few miles down here river channel in mountain behaves like confirmed system, whereas a moment it comes out in the plain, Terai in the south the percolation is maximum in the fan area and goes decreasing in the south up to 8 to 10 miles. The river starts gaining in the middle of the Terai to the Indo-Nepal boarder from the bank storage or leakage from the leaky aquifers of the ground water (Sharma, 1997).

Table - 9.1

INFILTRATION RATES IN THE DIFFERENT AREAS

S.N.	LOCATION/PLACE	TOP LAYER	25 cm FROM SURFACE
01	Parsauni Lumbini	8.8	2.8
02	Pharsatikar	3.2	4.5
03	Dogara	7.0	4.8
04	Jura	4	2.8
05	Parsauni S.W.	3.9	4.3
06	Semri	2.7	2.9

Source: Sharma, K.C., 1997: Rivers of Nepal, p. 57.

Table no. 9.1 indicates the minimum percolation rate is 3 mm per hour i.e. 72 mm per day of course, these rates are high as data belong to dry period whereas, these rates are high as data belong to dry period whereas, rate decreases as water table comes up in the rainy and in winter seasons. The data are based on soil and in the river bed the percolation rate is high in Bhabar zone (boulder) i.e. around 2 to 10 meters in dry season and nearly 1 meter in rainy season, in the middle part of Terai it is nearly 0.7 meter per day and lower part it is 0.07 meter per day.

This indicates that river traversing in Terai loses its discharge heavily. This is reason why in dry period most of river discharge becomes in significant. The percolation from the river can be calculated as follows:

RST= FxLxBxT/1000

RST=MCM Total percolation:

Where, F= average percolation rate in m/day

L= length of river in km.

B= width of river in M

T= length of time in days

In case of Tinau, Nov-May percolation is 0.4 m/day

RST=0.4x8x20x210/1000=13.4 McM

This gives an idea of percolation of the river. As long as infiltration capacity exceeds the rate at which rainfall strike the surface all incoming water will be infiltrated and non will runoff (Hortons, 1945).' 'The basic assumption is that is all infiltrated water delayed in its procession to stream channel because it percolates downward to the groundwater table and from there moves riverward and slope velocity at the ground water slope. Such flow would be a part of river flow but would enter the channel after direct the runoff accruing from the initiative storm was moves downstream' (Chorley, 1978). The same situation is prevails in the almost of all the area of Terai region of Sapt Kosi river basin. High infiltration rate are commonly absorbed in the Shivalik area as well as the pediment area of the middle Himalaya and upper Himalaya. The streams originating from the high altitudinal area are involves in down cutting and a deep valley develops but when these rivers downpours sediment is being deposited in the bottom of comparatively large size in which the running water when came to this point vanished away. It is common feature in the Terai region. The whole Bhabar zone formed by the old alluvium signifies by infiltration capacity. Overland flow transverse a surface as a broad shallow seed on in a small linear depression called rills. The flow is most common on slopes having limited will development and sparash vegeted cover. The intensity of infiltration varies in different part of Sapt Kosi catchment.

SUB SURFACE STORM FLOW AND SATURATED OVER LAND FLOW: The water surface in a river channel rises during times of high discharge. The relationship is predictable enough to provide the data for calibrating a rating curve by which discharge can be estimated directly from surface height alone. But at some particular discharge, depending on the channel geometry at the point of observation, a river will exceed the height of the channel banks, and overbank flow, or flooding, will occur. A flood is synonymous with discharge greater than bankfull channel capacity although many people carelessly refer to any high discharge as a flood. More detail study about flood will dealt in flood section of the chapter.

STUDY OF INDIVIDUAL RIVER HYDROGRAPH: To know the flow characteristics of the river, the hydrograph gives the behavior of the rivers at the spot. Most of rivers of mountainous region have a sharp discharge hydrograph during monsoon season (July-September) and comes very rapidly down to November and finally becomes low in the February. Then starts rising with the snow melt. Study of individual river hydrograph is given as follows:

BAGMATI: Bagmati at Gaurighat indicates that maximum rainfall is immediately reflected in the maximum flood. Rainfall and discharge relation is approximately 4 mm to 1 cusec. Mean discharge is maximum in August, lowest discharge occurs in the month of May and June. As there is no snow field in the nearby area, the minimum discharge is related is related in hottest part of the year. The minimum discharge is maximum on October that is the declining period of the rainfall. It goes on decreasing till it reaches to May.

Table - 9.2
MEAN DISCHARGE IN BAGMATI RIVER AT GARIGHAT

S.N.	MONTHS	TOTAL AMOUNT OF WATER ($x10^6m^3$)
01	November to December	4.2
02	December to January	2.6
03	January to February	1.4
04	February to March	1.2
05	March to April	0.9
06	April to May	0.6
07	May to June	0.3
	TOTAL	11.2

Source: Sharma, K.C., 1997: Rivers of Nepal, p. 107.

During the off season the total amount of minimum water is 11 x 10^6 m^3 available and minimum is 0.3 X 10^6m^3 in April to May. To get more one has to reserve. To balance the low flow to

Table - 9.3

AVERAGE MONTHLY AND YEARLY WATER DISCHARGE (in m³/s) OF BAGMATI RIVER (PANDHERA DOBHAN)

Year	Jan.	Feb.	Mar.	Apr.	May	Jun.	Jul.	Aug.	Sep.	Oct.	Nov.	Dec.	Year
1979	...	17.1	11.2	11.4	11.3	51	423	445	181	67.3	22.1	...	...
1980	...	...	11.1	9.31	12.9	123	314	338	172	68.1	23.6	...	...
1981	16.2	13.5	11.3	14.5	27.3	64.8	280	364	251	82.6	34.3	20.4	98.3
1982	16.8	18.8	12.8	15.8	12.4	122	323	249	282	51.9	30.6	22.7	96.5
1983	...	...	...	12.2	38.9	20.5	387	264	266	135	39.5	24.8	...
1984	20.4	14.7	9.38	9.56	33.5	166	538	300	987	79.1	31.5	22.3	184
1985	20	13.9	10.1	8.33	21.2	112	407	373	395	123	29	27.4	128
1986	...	...	...	...	...	...	...	...	...	...	...	...	...
1987	...	...	...	...	...	37.3	551	558	472	281	82.7	32.4	...
1988	20.8	14.9	17.4	15.3	50.2	140	461	755	291	64.9	39.7	28.1	158
1989	26.9	17.5	15.7	13.1	26.9	64.5	393	286	423	92.8	32.8	22.1	118
1990	18	18.4	18.8	17.4	52.1	131	407	484	364	231	90.4	41.8	156
1991	26.4	9.79	8.29	12.8	29.9	178	415	436	375	144	50	26.3	143
1992	18.4	15.6	10.7	8.76	16.5	92.2	316	356	261	180	41.8	24.4	112
1993	18.4	15.3	14.9	21.3	28.1	109	793	811	465	205	75.6	41.6	216
1994	29.5	21.3	13.7	11.8	27.1	167	301	266	272	52.6	24.8	16.6	100
1995	12.8	12	9.55	7.83	37	175	336	365	162	80.4	53.3	28.6	107
1996	24.7	17	13	10.4	15.3	205	525	458	358	178	57.1	30.5	158
1997	23.1	17.3	12.3	28.4	24.8	118	373	542	240	71	39.7	44.6	128
1998	25.4	21.1	21.5	23.9	42.9	77.4	839	654	208	89.8	39.2	20.7	172
1999	15.9	13.5	10.8	9.33	53.1	273	440	481	189	199	35	21.4	145

2000	18.9	16.4	14.7	14.4	44.5	109	322	599	169	49.7	29.4	20.3	117
2001	17.9	15.6	14.9	14.6	40.9	125	320	474	355	104	51.8	38.4	131
2002	32.5	29.5	24.6	26.1	56.6	107	897	235	195	71.8	30.3	18.7	144
2003	16.3	20.6	18.4	17.9	17.1	95.8	451	343	213	68.1	29.1	16.2	109
2004	14.5	12.1	10.3	11.9	22.4	78.4	676	100	142	109	40	25.5	103
2005	23.1	18.4	17	14.1	27.3	48.6	282	396	72	40.4	21.9	14.5	81.3
2006	11.5	11	14.1	10.6	25.4	131	169	158	406	51.6	20.8	16.2	85.5
Average:20.4		**16.5**	**13.9**	**14.3**	**30.6**	**116**	**442**	**411**	**302**	**110**	**40.6**	**25.8**	**130**

Source: DHM, 2014; Kathmandu, Nepal

the tune of 4.2 X 106m³ per month a reservoir is required to hold 18 X 103 m³ which means 1 km X 1 Km X 18m size of 3 reservoirs.

Data for discharge of the Bagmati river has been collected for Pandhera Dobhan stations for the 28 year which has been exhibited in table no. 9.3. It portrays that in all the succeeding years discharge has not remain equal the variability in stream discharge shows anomalies in the hydrographic aspects it indicates towards the decreasing rate of river mean water discharge. It could be possible by change in monsoon rainfall pattern in basin area. It also indicates the changing behavior of climate in the basin.

DUDHI KOSI: Likewise the other rivers the discharge in the Dudhi Kosi river becomes low in March and snow melt starts from April to June and peak is seen from July to August. The gap between maximum and minimum is small (Fig. no. 9.1).

The rate of mean discharge is highly fluctuating. It also indicates towards constantly increasing rate. It reveals that in all the succeeding years discharge has not remain equal the variability in stream discharge shows anomalies in the hydrographic aspects it indicates towards the vegetation degradation in the upper catchment area, climate change as well as high rate of glacier retreat in upper catchment area. The rate of discharge in each tributaries of Dudhi Kosi also

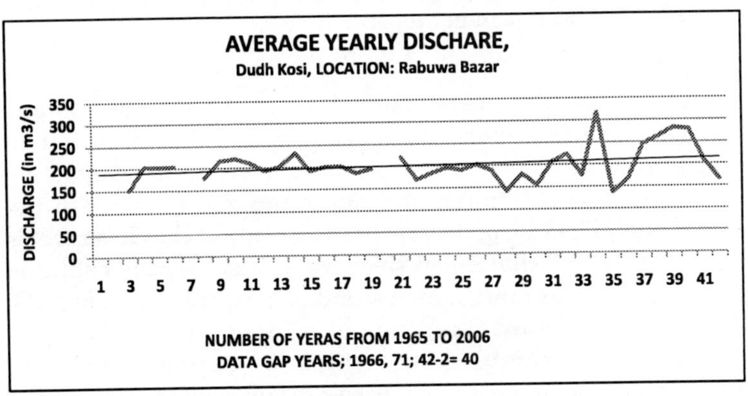

Source: DHM, Kathmandu, Nepal

Fig. 9.1

indicates towards the increase sediment load and sedimentation.

SUN KOSI: Sun Kosi becomes low in March and gradually rises in May by snow melt and reaches peak in July and slowly comes down. The hydrograph become flatter after December. The maximum and minimum gap is not much. The rate of decrease 5.1 cusec per day from October to January.

It is evident from Fig. no. 9.2 that during 37 years mean yearly water discharge in continuously decreasing in the river basin. There are several causes for the changing pattern on water discharge of the river. Among them changing behavior of monsoon rainfall could be the main reason.

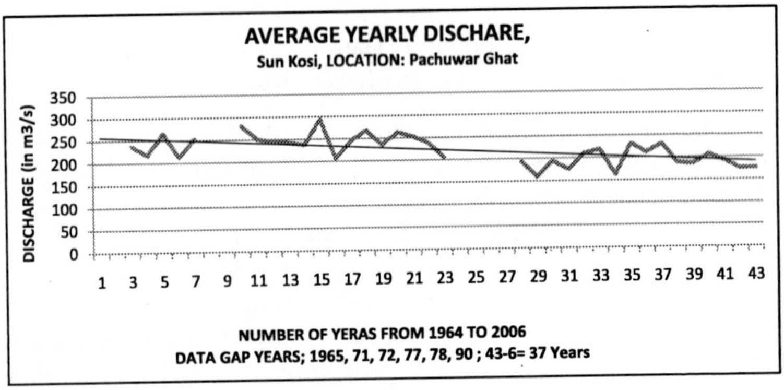

Source: DHM, Kathmandu, Nepal

Fig. 9.2

TAMUR: Tamur is one of main tributary of Sapt Kosi system. The river sharply comes down to low in March and snow melt helps up to April and then it shots to peak till the August by monsoon. The maximum and minimum gap is wider in the upper part. The rate of decrease is 4 cumec per day. It is reveals from Fig. no. 9.3 that mean river water discharge is continue increasing. It was maximum water discharge recorded in 1998; it was about 600 m^3/s. continuously increasing mean water discharge in Tamur river which is basically feed by Kanchenjonga's glaciers melt water in summer season.

ARUN: Arun river is the longest river in Sapt Kosi river basin

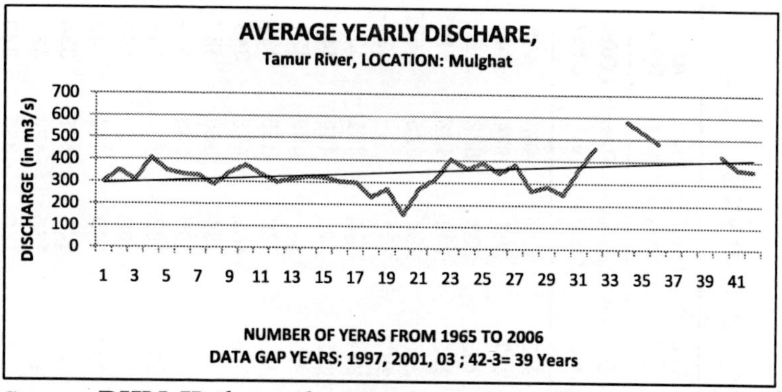

Source: DHM, Kathmandu, Nepal

Fig. 9.3

which also feeded by glacier melted water of Mt. Everest and other important peaks of Great Himalaya and northern basinal peaks of Tibet plateau.

It is evident from Fig. no. 9.4 that (19 years river data) mean water discharge since 1986 is decreasing year by year. Highest water discharge was recorded in 2000 year (more than 300 m^3/s), up to 2006 it decreased on less than 200 m^3/s.

A glance at Table no. 5.6 reveals the fact that

i. The maximum discharge in Sapt Kosi river basin occurs in the month from July to October.

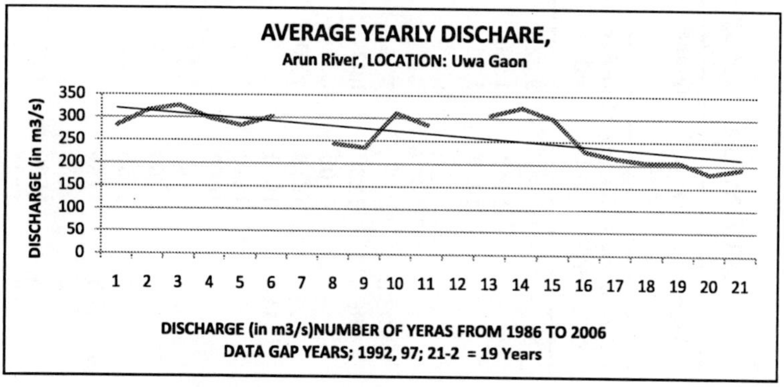

Source: DHM, Kathmandu, Nepal

Fig. 9.4

Table - 9.4
AVERAGE YEARLY WATER DISCHARGE (in m^3/s) IN SAPT KOSI RIVER

River	Bagmati	Sapta Koshi	Bagmati	Bhote Kosi	Tama Kosi	Khimti Khola	Balephi Khola	Tamur River	Sun	Dudh Kosi	Sun Kosi	Arun River
Station	(Pandhera Dobhan)	(Chatra)	(Sundari jal)	(Bara bise)	(Busti)	(Rasnalu Village)	(Jalbire)	(Mulghat)	(Pachuwar Ghat)	(Rabuwa Bazar)	(Khurk ot)	(Turke Ghat)
1976	-	-	2.22	70.6	124	27.3	42.8	299	242	193	463	405
1977	-	1480	1.13	73.4	145	21.9	47.6	313	237	203	474	445
1978	-	1660	1.32	81	162	28.8	53.2	323	293	233	622	465
1979	-	-	0.825	70.9	120	19.2	40.4	320	205	192	438	419
1980	-	-	1.2	82.8	149	31.4	41.7	302	245	201	525	568
1981	98.3	-	1.15	-	144	27.8	45.9	299	267	202	-	512
1982	96.5	1230	0.937	76.9	121	33.7	48.5	233	235	186	361	397
1983	-	1400	1.8	95.2	148	38.5	73.5	270	262	195	410	440
1984	184	1610	1.66	90.4	157	35.1	56	155	252		498	424
1985	128	1810	1.04	81.6	164	42	74.7	271	238	220	359	438
1986	-	1480	1.43		143	34.8	67	315	204	169	475	349
1987	-	1660	-	98.2	-	-	-	406	-	184	560	427
1988	158	1740	1.34	139	-	-	57.2	362	-	196	559	444
1989	118	1780	1.48	74.2	-	34.4	55.8	389	255	190	-	447
1990	156	1750	1.38	64.5	-	33.7	54.6	344		203	-	440
1991	143	1740	-	111	145	48.3	46.4	382	193	189	409	449
1992	112	1350	1.14	65.6		39.5	36.4	264	159	141	336	338
1993	216	1590	1.27	83.5	146	68.2	47.7	283	194	180	424	381
1994	100	1420	1.24	76.8	141	30.2	43	249	176	155	335	323

1995	107	1780	0.884	92.9	159	27.9	50.4	364	210	207	460	554
1996	158	1680	1.68	117	183	18.5	48.5	458	219	224	633	534
1997	128	1300	0.926	73.6	130	17.5	43.8	-	163	176	424	434
1998	172	-	1.16	92.8	178	24.7	71.4	579	231	317	463	553
1999	145	2070	1.4	-	142	19.6	64.4	532	212	137	663	487
2000	117	2030	1.13	-	159	19.3	56.4	483	230	167	690	500
2001	131	1760	1.35	-	146	24.4	49.9	-	189	246	-	466
2002	144	1650	1.53	-	-	18.2	53.6	469	186	263	564	424
2003	109	2040	1.13	-	-	17.6	52	-	206	282	-	544
2004	103	1450	1.1	-	132	20.8	54.6	424	193	279	532	-
2005	81.3	1240	1.09	65.6	122	19.2	58.3	365	175	209	532	456
2006	85.5	1310	1.44	65.9	145	19.4	56.2	358	176	166	553	466
MEAN	**130**	**1616**	**1.3**	**84.4**	**146**	**29**	**53**	**350**	**217**	**203.7**	**492**	**451**

Source: DHM, 2014; Kathmandu, Nepal

ii. The highest water discharge is found in the Sapt Kosi river (in Nepal) means is 1616 m^3/s. A record of water discharge taken at Chatra in Nepal, although the discharge data obtain from DHM, Kathmandu reveals that the highest record of maximum discharge has been recorded in the year of 1999. A record of 31 years shows that the amounts of dishcharge have been increased in the succeeding years.

iii. It is also observed that the stream having higher rate of discharge are Sun Kosi, Sapt Kosi, Arun and Dudhi Kosi.

iv. The remarkable features of discharge are that the amount of the discharge also increases in the summer season. It is because of the supply of water through the streams originating from upper reaches where the area is completely frozen.

v. Melting of ice which supplies water in the summer season, sedimentary record indicates high fluctuation in the monsoon conditions.

GROUND WATER: Ground water is an important hydrologic measure which helps in the low flow of the river. Ground water refers subsurface water which occurs below the surface of the earth and main source of it is infiltration (Brown, 2002). A ground water level indicates the atmospheric pressure of the

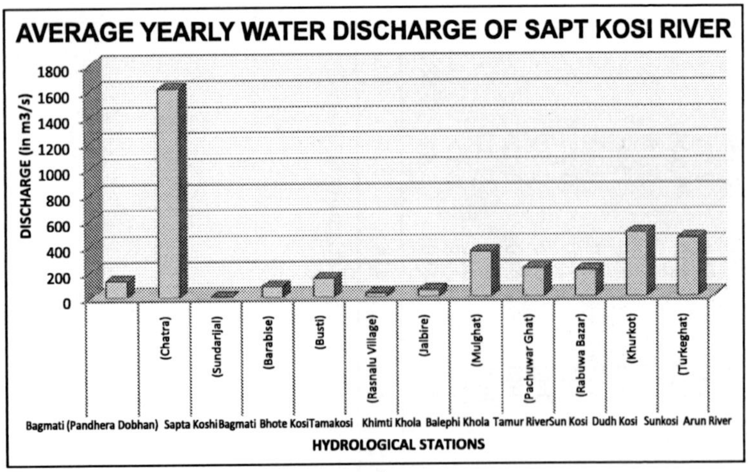

Source : DHM, Kathmandu, Nepal

Fig. 9.5

aquifer. Stream flow variation are closely related to the ground water levels where as the other causes are meteorological and tidal phenomena, urbanisation, earthquakes and external loads and last but not least the subsidence of the earth surface (Rizvi, 2008). Seasonal pattern of fluctuation is due to influences such as rainfall and irrigation pumping that follow the well drained cycle. The ground water potential in Sapt Kosi river basin is highly variable. The upper and middle catchment area of the basin has comparatively meager ground water potential where as the Terai and plain area having rich potential of ground water level. The rivers are generally act elevation of 200 feet in midland whereas, the rock section exposed above river channel is on the average of 4,000 feet or 1,300 meters. If one multiplies total drainage area with the thickness of rocks the amount of cubic meter comes and the rock has to take the total amount of flow water and ratio given the porosity of the average terrain (Table no. 9.6).

Table - 9.5
GROUNDWATER LEVELS AND POROSITY PERCENTAGE OF KOSI RIVER BASIN

S.N.	Rivers	Catchment Area	Height	Total Cubic Kilometer of Water	Porosity Percentage
		(sq. km)	(x1300m)		(%)
01	Arun	34,660	25 X 10^6	3038	0.012
02	Tamur	6,741	9 X 10^6	776	0.015
03	Sun Kosi	16,743	16 X 10^6	2263	0.016
04	Bagmati	13,589	6 X 10^6	1240	0.023

Source: Sharma, C.K., 1983, Rivers of Nepal, p. 104.

The hydrological behavior of the Sapt Kosi river basin though the groundwater may point out as follows:

i. The groundwater potential in the upper catchment area of the basin marked by permafrost, during six to seven months where as two to three months as a aquifers some places and some where medium to low.

ii. The groundwater is highly fluctuating and variable, even in the month of rainy season particularly during the rainy months rate of infiltration marked by very slow.

iii. The lower most catchment area of the basin characterised by deep sedimentary deposition having highly porous which allows high infiltration of water

So, the whole terrain region and plain means up to the mouth of Sapt Kosi river marked by highly saturated soil. Low infiltration rate of water is also responsible for flood water logging and marshy land of a long period.

FLOOD IN KOSI RIVER: The geomorphic effectiveness of large flood has been an important issue in the hydrology study. Flood is an important disaster creating a huge loss of human life and cultural environment. The flood in Sapt Kosi river basin is a chronic and area is very prone to from the time immemorial. The flood in a catchment depends upon the characteristics of the catchment, rainfall and antecedent conditions. The history of Kosi river is much older than the great Himalaya. Generally, floods are considered to be associated with rivers and people conceive flood as the outcome of accumulation of huge volume of water coming out of the rivers through overtopping of river bank during peak discharge period. It is a comparent of hydrological cycle of a drainage basin (Singh, S., 2014). Floods are associated with the alluvial rivers draining extensive alluvial and flood plains. Sapt Kosi river is the result of land born flood, landslide, dam-failure flash flood, rapid onset flood, heavy incessant flood and blowing of natural water in the upper catchment area etc. Sapt Kosi has average water flow of 2,166 m^3/s (76,500 $feet^3/s$). During flood, it increases to as much as 18 times as average. The greatest recorded flood was 24,200 m^3/s (8,50,000 $feet^3/s$). Extensive soil erosion and landslide in the upper catchment area have resulted in the silt yield of Sapt Kosi of about 19 m^3/hec/year. The major tributaries as Arun, Bhote Kosi, Dudhi Kosi, tamur carried huge amount of coarse silt in proportion of its total sediment load. The river is able to transport heavy sediment down to the steep gradient and narrow gorges of mountains and foot hills, where the gradient is at least 10 m/km. In the plain beyond Chatra theier is 1 m/km gradient where as in plain region of Bihar there is 0.04 m fall of gradient in per one km.

Table - 9.6
OCCURANCES OF FLOOD HAZARD IN KOSI RIVER BASIN

River Name	Site Name	Highest Flood Level/Year	Danger Level Level	1996 Date Level	1997 Date Level	1998 Date Level	1999 Date Level	2000 Date Level	2001 Date Level	2002 Date Level	2003 Date Level	2004 Date Level	2005
Sapt Kosi	Barah Kshetra	132.18/ 1968	120.30	19/08- 123.09	18/08- 123.70	31/08- 123.89	26/08- 124.65	20/08- 125.53	31/07- 121.32	24/07- 124.05	10/07- 122.81	11/07- 122.51	27/08- 122.36
	Birpur	76.85/ 1968	74.70	15/07- 75.13	16/09- 74.61	06/09- 75.59	03/08- 75.34	03/08- 75.59	16/07- 75.78	24/07- 75.95	10/07- 75.89	11/07- 75.28	27/08- 74.98
	Basua	48.76/ 1996	47.75	21/07- 48.60	14/08- 48.20	06/08- 48.63	27/08- 48.58	03/08- 48.64	31/07- 47.89	25/07- 48.24	10/07- 48.61	12/07- 48.58	27/08- 48.30
	Baltara	36.40/ 1987	33.85	23/07- 35.59	20/08- 35.14	24/08- 35.77	29/08- 35.67	07/08- 35.58	31/07- 34.51	30/07- 35.96	22/07- 34.89	14/07- 35.97	30/08- 34.98
	Kursela	31.86/ 1998	30.00	03/09- 31.56	16/08- 30.58	03/09- 31.89	27/09- 31.50	12/09- 31.10	31/07- 38.45	27/07- 30.48	27/07- 29.76	24/07- 30.04	30/08- 31.10
	Jhajharpur	52.73/ 1987	50.00	21/07- 52.21	13/07- 51.57	28/07- 52.53	04/07- 51.69	02/08- 58.48	31/07- 50.79	23/07- 52.53	10/07- 51.73	10/07- 53.00	27/08- 51.42

Source: Disaster management dept., Govt. of Bihar (http://wrd.bih.nic.in).

A glance at Table no. 9.7 portrays that occurrences of flood is a recurring phenomena in Sapt Kosi river basin. Though its effectiveness varies in the different years. All the villages in the lower catchment area of Sapt Kosi river basin flooded during the monsoon period. This hydrologic aspect has definitely damage the basin by shifting its channel, expansion of flood water and post flood phenomena. It is also obvious from the Table no. 9.8, that more than 45 blocks of the basin particularly blocks of Muzaffarpur, Sitamarhi, sheohar, Darbhanga, Madhubani, Samastipur, Begusarai, Khagaria, Supaul and Madhubani. 6 blocks of Darbhanga, six blocks of Khagaria remains more than 15 to one month under flood. Flood in Sapt Kosi river basin is highly geomorphic as the melt water along with monsoon rainwater flows down rapidly owing to high steep gradient of the middle and upper catchment area which ultimately result into flood into lower catchment. The embankment made for the protection becomes failure when the accumulated water out breaks forming flood in the lower part of the basin.

Table - 9.7

NUMBER OF BLOCKS AFFECTED BY FLOOD

(1987 - 2011)

S.N.	Name of Districts	Number of Blocks Effected out of Last 25 years16 years and more			
		11 years	15 years 1 years	6 years 5 years	10 years
01	Muzaffarpur	5	3	2	6
02	Sitamarhi	6	6	6	0
03	Sheohar	1	2	1	1
04	Madhubani	2	8	9	5
05	Darbhanga	6	6	6	2
06	Samstipur	3	3	6	10
07	Begusarai	-	3	4	9
08	Khagaria	6	1	-	-
09	Saharsa	5	1	-	3
10	Supaul	5	1	1	5
11	Madhepura	2	-	6	4

Source : Disaster management dept., Govt. of Bihar (http://mrd.bih.nic.in)

The recurring flood has been responsible for loss of public property since long period. A loss of public property (cultural landscape) owing to flood has been shown in Table no. 9.9 and Fig. no. 9.6. It clearly reveals that the flood from 2001 onwards has been responsible for crucial damages in the basin area.

PALAEOCHANNELS OF SAPT KOSI RIVER: Palaeohydrology is the study of the flood processes operated in the past and their hydraulic implications. Fluvial palaeohydrology is that branch of palaeohydrology which deals with erosion, deposition and the characteristics of former channels (Cheetam, 1976). Palaeohydrology treats the phenomena of occurrence of water in the atmosphere, its distribution and the composition of the surface of the ground and underground but has reference to the past (Schumm, 1977). The palaeohydrology restricted to that portion of hydrologic cycle which incloses the movement of water over the surface of the earth, because runoff and the accompanying sediment load are major important in determining the non glacial erosion and depositional features of the quarternery. Thus it involves consideration of quality and quantity of the sediment move through the paleochannels (Gragrey, 2005). Study of palaeohydrology involves palaeoclimatology, palaeogeomorphology, palaeohydraulics and palaeopedology.

Sapt Kosi river is noted for continuously migration. It has been counted that the main channel has been migrated more than 130 km westward across the surface of the mega fan over the last two centuries. Examination of a number of old maps published between 1760 and 1960 shows that during most of this period Sapt Kosi River occupied a position slightly east of the megafan axis. The apparent channel movement shown in these maps is oscillating in nature and not unidirectional. Instead of encountering deposits left behind by a sweeping braided Kosi-like stream, a preliminary study of the uppermost 25 m succession in the north-central part of the megafan reveals overwhelming dominance of meandering stream deposits. Assuming the existing notion of Kosi River migration, the rate of deposition averaged over 100-200 years for the uppermost 510 m of megafan deposits, works out to be unusually high (>50 mm/y). The existing facies model for the uppermost 810 m of

Table - 9.8
LOSS OF PUBLIC PROPERTY SINCE YEAR 2000

Year	Number of Effected Dis-trict	Bloc-ks	Vill-ages	Human	Lack (In Animal)	Area (In lack HA) Agri	Non-Agri	Total	Crop ped	Crop Damaged (RS/ Lack)	House Damaged Total	Value (In Rs/ Lack)	Public Property Damaged (In Rs)	Deaths Human	Animal
2011 (P)	24	154	3588	64.171	5.98	2.642	0.639	3.001	1.279	5627.00	28067	12874.10	3578.6	143	33
2010 (P)	8	41	489	7.22	0.56	0.631	1.112	1.743	0.03	202.45	8733	479.26	59.2	28	2
2009	16	91	1546	22.03	1.346	1.71	9.339	11.05	0.475	2182.57	7674	528.15	530.10	97	2
2008	18	116	2585	49.952	12.166	6.405	2.12	8.824	3.672	3420.25	297916	8451.40	9771.96	258	878
2007	22	269	18832	244.42	27.13	13.323	5.51	18.833	10.603	7683782	784328	83144.52	64241.52	1287	2423
2006	14	63	959	10.89	0.1	1.52	0.297	1.81	0.87	706.63	18637	1225.03	8456.17	36	31
2005	12	81	1464	21.04	5.35	3.343	1.261	4.6	1.35	1164.50	5538	382.79	305	58	4
2004	20	211	9346	212.99	86.86	20.99	6.01	27.00	13.99	52205.64	929773	75809.51	103049.60	885	3272
2003	24	172	5077	76.02	11.96	9.943	5.14	15.08	6.10	6266.13	45262	2032.10	1035.16	251	108
2002	25	6	8318	160.18	52.51	14.45	5.244	19.69	9.4	51149.61	419014	52621.51	40892.19	489	1450
2001	22	194	6405	90.91	11.7	9.042	2.91	11.95	6.5	26721.79	222074	17228.44	18353.78	231	565

Source : Disaster management dept., Govt. of Bihar (http://mrd.bih.nic.in)

the megafan deposits also appears untenable (Chakraborty, Kar, Ghosh & Basu, 2010). The estimation of palaeo discharge and velocity and gravel bed rivers has been computed applying the following equations.

PALAEO VELOCITY DETERMINATION: With a standard error of 12%. Here slope was obtained from topographic maps. On the basis of the above equation palaeo channels of Sapt Kosi river have been determined. The velocity of the palaeo channels has been measured to 2.6 m^3/s which shows the channels have been exaggerated for extent from the original flows.

PALAEO DISCHARGE DETERMINATION: Williams (1988) used 233 river cross sections from a variety of environments and developed the empirical equation for bankful discharge Q_b. The palaeo discharge of Sapt Kosi river basin has been determined using the above equation. On that basis the value of the discharge has been obtained to which indications that from the beginning the tendency of the discharge is highly fluctuative the discharge velocity varies to such an extent that in the summer season discharge velocity remains very low throughout the rainy season. It is very high in the cold weather season discharge remain constant.

PALAEO VELOCITY AND PALAEO DISHCHARGE IN GRAVEL-BED RIVERS: Maizels (1983) has stated that in the case of gravel-bed rivers it is preferable to assume that critical

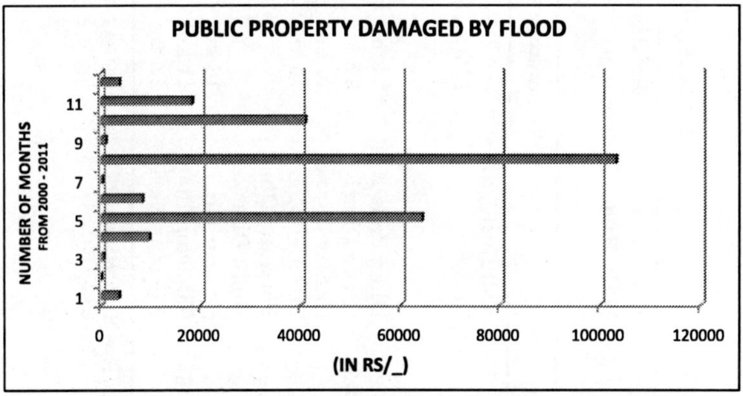

Source : Disaster management dept., Govt. of Bihar (website)

Fig. 9.6

Table - 9.9
PALAEOCHANNELS OF KOSI RIVER BASIN

Year	Name of the Main River Course	Shifting	Period of Shift	Average Rate of Shift	Remarks
		in km	(in Year)	(km/Year)	
1731	Kali Kosi	-	-	-	-
1770	Livari	11 km	39	0.28	Period of shift Has been taken as midpoint to span of Live existence of two Successive courses. Period of rapid shift.
1807-1839	Dhamdaha Kosi	8 km	53	0.15	
1840-1873	Kiran Dhar	6.5 km	33	0.19	
1875-	Sursar Dhar	11 km	27	0.40	
18921922	Mirchai Dhar	16 km	24	0.66	
19221936	Parwane Dhar, Belhi Dhar & Dhusar Dhar	16 km	22	0.72	-
19361946	Tilawae Dhar	11 km	7	1.57	Period of rapid shift.
1954	Flowing Kosi	27 km	18	1.80	-

Source: The Kosi River, its morophology and mechanics in retrospect & prospect" by C.W.C in April 1982 conditions prevail at bankful stage. The palaeo velocity and palaeo discharge of a river basin are indicative of the erosive power of the stream. It has been continuously determined by the using the above formula. By the above equation empirical discharge relating to mean annual discharge or bankful discharge to the channel dimension or channel form parameters is determined. The empirical drainage area discharge relation and regime type relations in which bankful discharge is relted to width or depth combinedly computed.

THE PALAEO HYDOLOGIC STUDY OF SAPT KOSI RIVER BASIN: The palaeo hydrologic study of the Kosi river basin related to establishing palaeo flood records. The method used for analysing and measuring is of slack water deposits (SWG) are five grained sand and silt deposit that falls rapidly out of suspension during large floods in pritceted area where the fall velocity is markedly reduced such deposit takes place in stable bed rock, canyon expansion and contractions, bank flood tributaries, meander bends, caves and abrupt channels (Kothyari, & Garde, 1991). The palaeo channels of Sapt Kosi river basin with related data have been exhibited in Table no. 9.10 and Fig. no. 9.8.

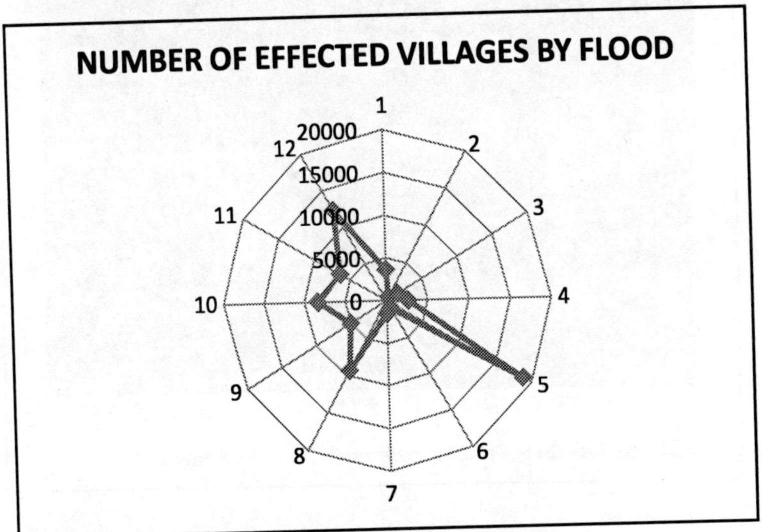

Source : Disaster management dept., Govt. of Bihar (Website)

Fig. 9.7

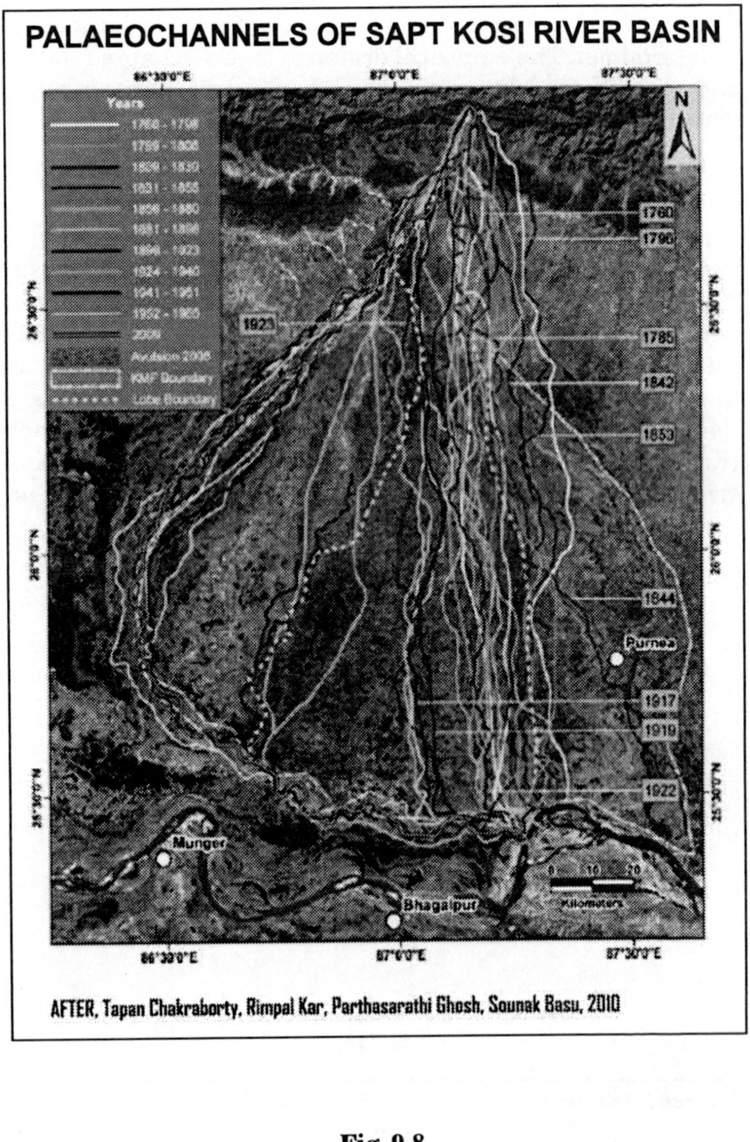
PALAEOCHANNELS OF SAPT KOSI RIVER BASIN
N
1760
1796
1923
1785
1842
1853
1844
Purnea
1917
1919
1922
Munger
Bhagalpur
86°30'0"E
87°0'0"E
87°30'0"E
AFTER, Tapan Chakraborty, Rimpal Kar, Parthasarathi Ghosh, Sounak Basu, 2010

Fig. 9.8

The palaeochannel study as it reveals from the Table no. 9.10 that in the earlier time Sapt Kosi joined the river Mahananda through the present course of the river Parnar near Araria in Purnea district. This view is supported by the present of long stretch of depression varying in width from 30 m to 60 m passing from Forbesganj side towards Araria. However, since 1736, the river has shifted towards west through 112 km in 223 years, in this processes it has deposited sand over 7,680 sq. km land in Bihar, 1,280 sq. km in Nepal making the land almost infertile. The positions of Sapt Kosi main channel flow during different years are shown in Fig. no. 9.8. The shifting has been always towards the west and the average shifting rate has varied from 0.19 km/year to 1.8 km/year with an average rate of 0.478 km/year. At Purnia to Belhi. The river has shifted westward by abounding its old channels, since Sapt Kosi and its tributaries streams covers high amount of sediment load, much of the sediment deposit during the recession of the flood their by choking mouth of the same channel. As a result, during the next flood the river activates a new channel which gets developed during the passage of time.

CHAPTER – 10

SEDIMENTATION

The sedimentary environment is the complex of physical, chemical and biological conditions under which a sediment accumulates. Sedimentation operate within a framework controlled by the physiographic setting of the environment, by relations of land and sea, and by the action of dominant geological agents, such as streams, wind, waves, or currents (Krumbein, & Sloss, 1953). Sedimentary environments vary widely in their persistence through time, in the size of the area occupied by them, and in the uniformity or variability of their conditions. An ox-bow lake on a flood plain, on the other hand, may be small and ephemeral, changing to a bog within the span of a few years.

In any sedimentary environment, the sediment is derived from some source, it is carried from the source to the point of deposition by some agents, and the agent is energized in the source area, and depositional landforms are developed in the source area, and depositional landforms and sedimentary deposits are formed in the sedimentary environment. Under suitable conditions, organisms in the environment may greatly affect the accumulating sediments (Krumbein, & Sloss, 1953). Therefore, sedimentation is the tendency for particles in suspension to settle out of the fluid in which they are entrained, and come to rest against a barrier. This is due to their motion through the fluid in response to the forces acting on them: these forces can be due to gravity, centrifugal acceleration or electromagnetism. In geology sedimentation is often used as the polar opposite of erosion, i.e., the terminal end of sediment transport. In that sense it includes the termination of transport by saltation or true bedload transport. Settling is the falling of suspended particles through the liquid, whereas sedimentation is the termination of the settling process. Sediment in the rivers are two types that is suspended sediment and bed load. Suspended sediments are fine and are mixed with water which is cause of silting and bed load are reason for channel shifting. Mass of sediment increases with landslide, mass movement, deforestation, rainfall etc.

ELEMENT AND FACTORS OF THE SEDIMENTARY ENVIRONMENT: The study of any sedimentary environment, accordingly, includes four basic elements, which are the materials of the environment, the boundary conditions, the energy of the system, and the biological complex. Environmental factors are specific aspects of the environmental elements.

MATERIAL: The material of the environment includes the medium of sedimentation (air, fresh water, sea water and glacial ice) and the texture and composition of material in transit or being deposited. In sea water, the medium itself contains dissolved salts and gases, which are included among environmental factors. In case of Kosi river basin the main material is earth materials. The river basin obtain earth materials from landslides, erosion of valley surfaces in both form vertical and horizontal and material as a form of solvent in running water.

BOUNDARY CONDITIONS: The boundary conditions of the environment include the geometrical features of the environment, such as depth of water, shape of a bay, configuration of the depositional surface, and others. The boundary conditions set the depositional surface, and others. The boundary conditions set the geologic or physiographic framework of the environment. This element works dominantly in Kosi river basin. The lofty mountains of Himalaya and Tibet plateau plays role as aggradational forces where as in Terai and plain region act like depositional. It occurs only due to boundary conditions of the river basin.

ENERGY: The energy of the system includes such diverse factors as the kinetic energy of wind, the turbulent energy of waves and currents, or the thermal energy in a body of water. Part of the available energy may not be utilized, such as the energy of waves passing over a deep-water environment. If the bottom is not disturbed, the net effect of the wave energy on the sediment may be negligible. It is the energy which acts upon the material may be negligible. It is the energy which acts upon the material, or is dissipated in the environment by friction or other factors, that is important in sedimentation. The height (slope) of the river basin in upper catchment area plays dominant role for kinetic energy in water masses due to gravitational force of the earth surface where as in lower region of the basin due to low gradient loss of energy in the form of about stable water bodies.

BIOLOGICAL FEATURES: The fourth element of the sedimentary environment is the biological complexities. In some environment, organisms are the principal agent of sediment accumulation. Coal, coquina, and crinoidal limestone are example. In the upper catchment areas of the basin that vegetation bodies also flows in the water channel and influences in some quantities during sedimentation processes.

Table - 10.1

SEDIMENTATION IN SAPT KOSI RIVER BASIN

S.N.	Rivers	Catchment Area (Sq. km.)	Sediments ($10^6 m^3$)	Sediment Load (m^3/sq. km)
1.	Sunkosi	19230	54.2x	2818
2.	Arun	36533	34.6x	947
3.	Tamur	5900	29.6x	5016
Total		61663	116.4x	8781

Source: Sharma, C.K., 1983, Rivers of Nepal.

SEDIMENTATION IN SAPT KOSI RIVER BASIN: River systems of part of the Himalayan foreland, northern Bihar plains, India, are described in terms of their channel morphology, hydrology and suspended sediment characteristics (Sinha & Friend, 1994). The present-day Indo-Gangetic plains are not only one of the world's largest areas of Quaternary alluvial sedimentation, they also form the upper surface of one of the largest, still actively subsiding, foreland basins (Parkash & Kumar, 1991). The part of the Kosi River valley in India is an alluvial plain, an important part of the Ganges Plain (Chen, Hu, Deng, Khanal, Zhu, and Han, 2013). Huge sedimented mega fan

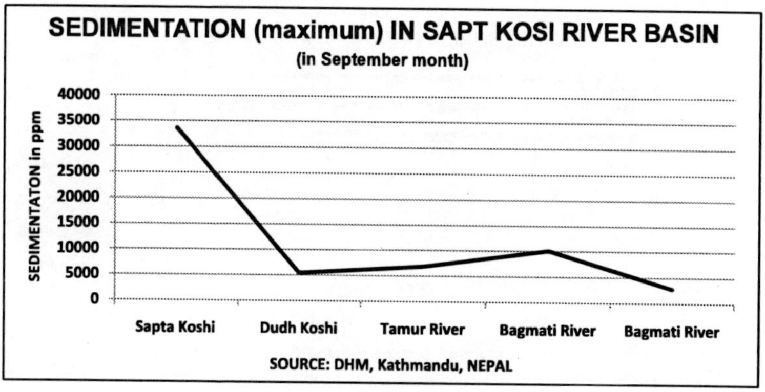

Fig. 10.1

is found in this part of Sapt Kosi river basin in a triangular shape. Behaviour of Sapt Kosi River is inextricably interlinked with the geomorphology of Sapt Kosi mega fan and its formative mechanism. Sedimentation process played a dominant role in the shaping of Sapt Kosi river basin particularly in the lower catchment area of Bihar plain.

Landslide and soil erosion is the basic key for the obtaining sediment materials in the basin. In which Tamur is the major source of sediments, it carries 5,016 m^3sq. km sediment load. But in scenes of total sediment, it is Sun Kosi river (54.2 X $10^5 m^3$) in middle Himalaya of Nepal carries highest amount of

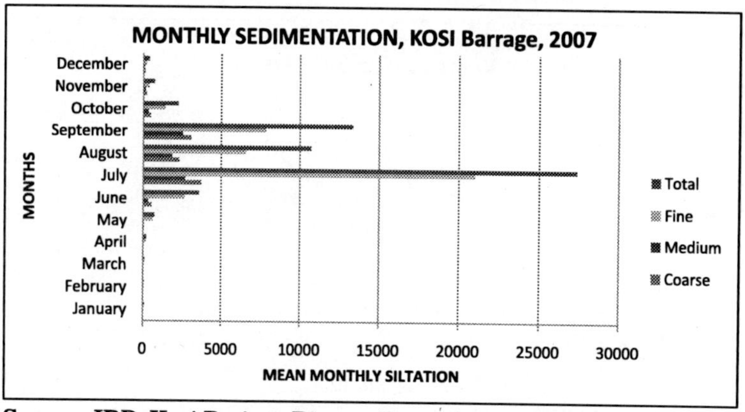

Source: IRD, Kosi Project, Birpur, (Supaul), Barrage site.

Fig. 10.2

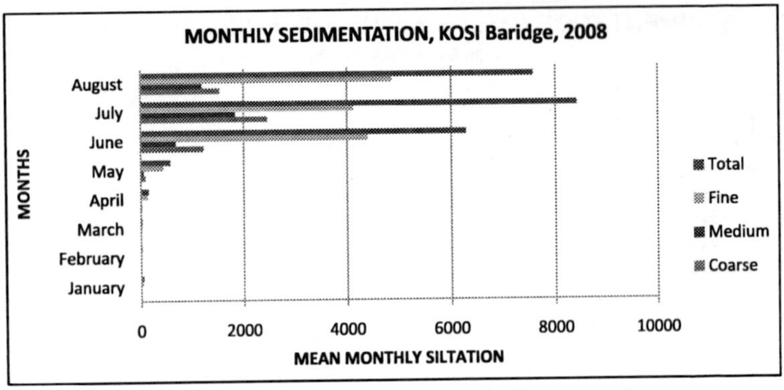

Source: IRD, Kosi Project, Birpur, (Supaul), Barrage site.

Fig. 10.3

sediment. These can be clear from the following data.

In the above Fig. no. 10.1, it is evident that Sapt Kosi carries highest amount of sediment where as Bagmati in second place, Arun river contributes lowest amount of sediment in Sapt Kosi river basin.

Monthly sedimentation is shown by Fig. no. 10.2 and 10.3. In these figures monthly sedimentation is high during rainy months of June, July, August and September that is 4,000, 27,000, 11,000 and 13,000 siltation respectively where as in 2008 year total siltation was higher in July month but it was only 8,500. That's mean comparatively it is very low quantity of siltation

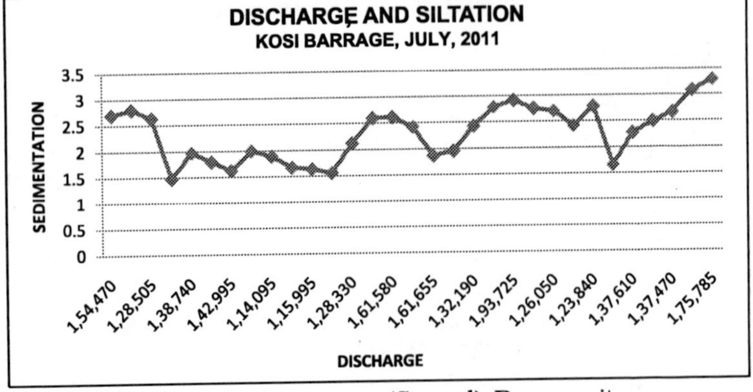

Source: IRD, Kosi Project, Birpur, (Supaul), Barrage site.

Fig. 10.4

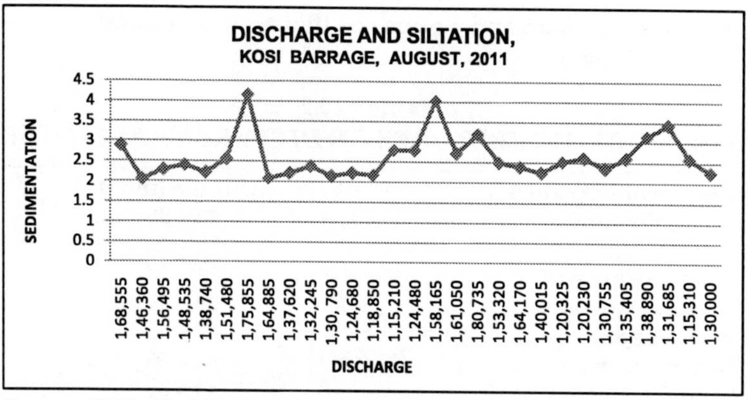

Source: IRD, Kosi Project, Birpur, (Supaul), Barrage site.

Fig. 10.5

occurred than year 2007. But another hand siltation of medium and coarse material was more released during 2008 year in compare than year 2007.

In Fig. no. 10.4, 10.5, and 10.6 there is sited graph of relation between discharge and siltation on the basis of data obtained from Kosi barrage, Supaul. There is absolutely close relationship between sedimentation and volume of discharge in the river channel. Suppose that during year 2011 we got that when discharge goes more than 2.6 lack of discharge than siltation also goes up upto 4.5 ppm which is comparatively higher among those three years. During month of August of 2011 twice time

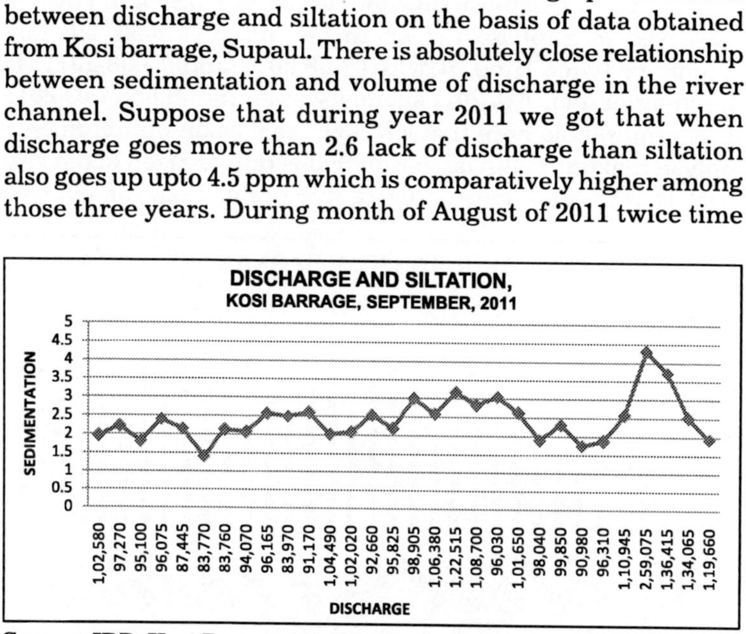

Source: IRD, Kosi Project, Birpur, (Supaul), Barrage site.

Fig. 10.6

discharge goes up and we can see that both time siltation goes on record level.

Table - 10.2

GEOMORPHIC FEATURES ON DIFFERENT ALTITUDE

Geomorphic Feature	Altitude (in m)
Active Channels	33 – 35
Channel Bars	35 – 37
Inactive Channels	37 – 39
Crevasse Deposits	39 – 41
Water-logged areas	41 – 43
Oxbow-lakes	43 – 45
Sand Patches	45 – 47
Flood Plain	47 – 49

Source: Calculated through Google earth images and topomaps

In Fig. no. 10.7, 10.8. and 10.9 the date wise monthly silt flow through Kosi barrage is sited. In Fig. no. 10.9 we can see that monthly variation in silt flow through Kosi barrage is found. 31.07.2011 was highest silt flow day; it was 3.4 ppm silt flow occurred in Sapt Kosi river. In Fig. no. 6.17 peak silt flow occurred twice time, it was 07.08.2011 and 15.08.2011. In September month of 2011 silt flow was optimum on 27.09.2011. It is remarkable here that silt flow reaches on optimum level only in maximum fluvial season that is rainy season of the year that is why we choose only rainy season data for analysis

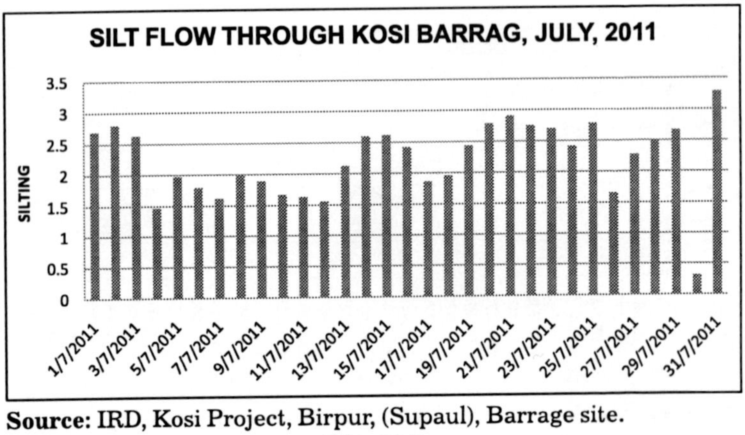

Source: IRD, Kosi Project, Birpur, (Supaul), Barrage site.

Fig. 10.7

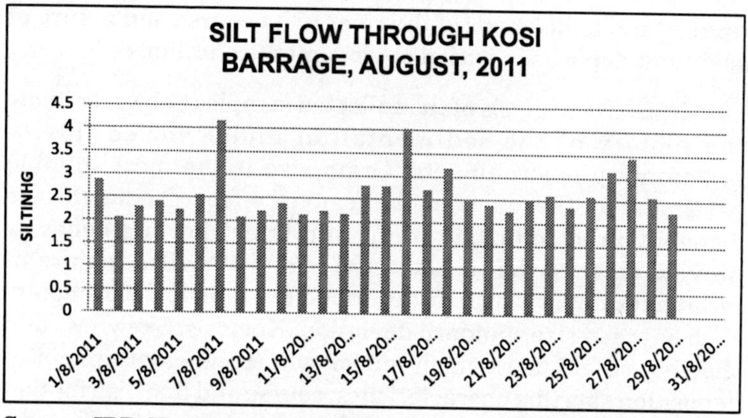

Source: IRD, Kosi Project, Birpur, (Supaul), Barrage site.

Fig. 10.8

purpose.

THE GRAIN SIZE ANALYSIS: The grain size analysis of the catchment samples from different sites help us to know the history of sedimentation under varied fluvio-sedimetary environment. Grain size is the most suitable parameter for reflecting the regional effect that sediment has to undergo. Sapt Kosi river basin is an important north bank tributary of the Ganga river system of northern India. As the river flowing through varied topography lithology and slope as well as socio-economic conditions, the study of the grain size helps to

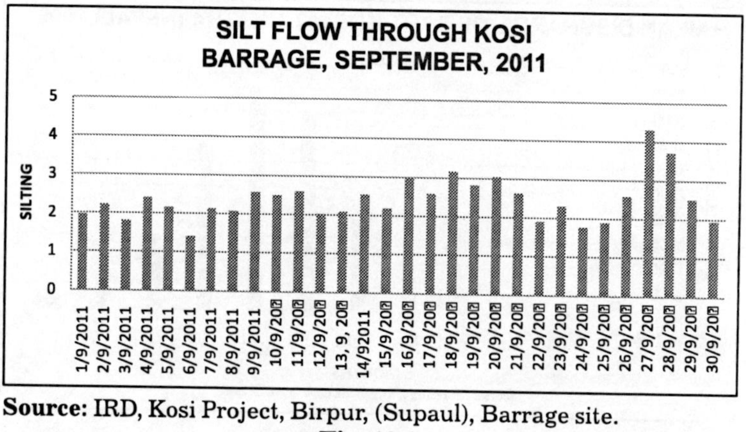

Source: IRD, Kosi Project, Birpur, (Supaul), Barrage site.

Fig. 10.9

understand to behavior of the river in its course and nature of sediment deposits in varied environment conditions.

Grain size analysis of the collected samples help us to know the history of the sedimentation under varied fluvio-sediemntary environments. Grain size is the most suitable parameter for the reflecting the regional effect of sediment size distribution of sediment grains and their shorting behavior provide close to the formation of sediments. The texture of sediment sample is usually described by statistical parameters such as mean size standard deviation, Kurtosis, Skewness and shorting co-efficient which intern reflect the environment of deposition shorting characteristics, course and fine distribution of sediments as well as distance of transportation of sediments. 'The present-day Indo-Gangetic plains are not only one of the world's largest areas of Quaternary alluvial sedimentation, they also form the upper surface of one of the largest, still actively subsiding, foreland basins (Parkash, B. & Kumar, S., 1991).' 'The part of the Kosi River valley in India is an alluvial plain, an important part of the Ganges Plain (Chen, N. Sh., 2013).'

For systematic mechanical analysis of the sediments sample have been collected from the different sample sites of Sapt Kosi river basin. Out of the samples are of bank material, bed materials, sand bar materials, which have been presented in the form of histogram, frequency distributions, the statistical

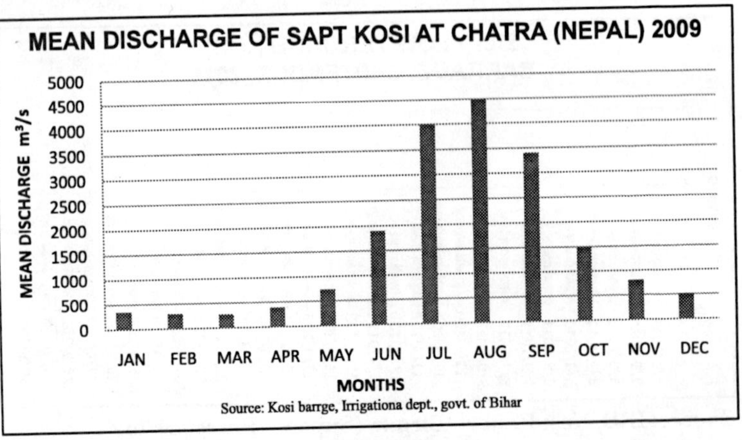

Fig. 10.10

measures are estimated from the curve so drawn.

BANK SEDIMENT ANALYSIS: The Bank sediments numbering 32 samples collected from 8 cross sections have been analyzed in terms of their compositional distributional and textural characteristics. The grain size parameters estimated from the collected bank sediment samples of each of the cross sections are interpreted in their inherent characteristics. The most commonly used size parameters like grain size class, diameter, weight percentage, mean size, standard deviation; kurtosis, skewness and sorting co-efficient have been collected and interpreted. At sample size H the bank sediments of size 0.250 to 0.177 mm represent highest weight percentage of 48.82, alternatively lowest weight percentage composed to 0.125 to 0.062 mm of sediment size (analyzed and shown in Table no. 6.3) the mean size of bank sediment of this size is 0.24 mm which means that the sediments belong to the fine sand category on the average. The value of standard deviation, kurtosis, skewness and shorting co-efficient are 0.46, 1.11, -0.75 and 1.12 respectively (Table no. 10.3) which suggests vary well shorted, moderately peaked, negatively sediment distribution at this site. At all the site, size of bank sediment varies, it maximum at A site that is 2 mm, where as the lowest weight percentage is observed in the range 0.125 to 0.062 mm of sediment. The mean size of bank sediment is 0.27 mm suggesting fine types of sediment on an average. The values of (site E) standard deviation, kurtosis, skeweness and sorting coefficient are 1.23, 0.77, -0.4 and 2.95 respectively which indicates that the sediments are moderately sorted and negatively skewed. The sand size of sediments at site D is 0.35 mm which means that sediment are fine sand the highest weight percentage is rendered within the sediment size of 3.54 0.250 mm. On the other hand the sediment of 0.0078 mm to 0.0039 mm size class account for 0.08%. The calculated statistical values of standard deviation, kurtosis, skewness and shorting coefficient are 0.94, 0.86, -0.55 and 1.86 respectively which reveals that sediments are well sorted. Likewise the statistical parameters of grain size distribution of bank sediments have been asserted in Table no. 10.4. There is, however, a general downstream decrease in grain size of bank material a set of eight cumulative curves for bank material is

Table - 10.3

BANK SEDIMENT SIZE DISTRIBUTION AT SITE A-H

Dia. In f	Dia. In mm	CUMULATIVE WEIGHT PERCENT							
		A	B	C	D	E	F	G	H
-1	2	17.60	11.32	3.67	7.11	12.64	0.43	5.09	0.56
0	1	20.12	13.75	6.33	10.34	21.67	1.05	5.65	0.87
1	0.500	41.45	30.67	17.80	25.33	44.32	2.54	7.35	2.95
1.5	0.354	53.75	43.60	31.56	35.24	46.54	9.56	8.58	13.66
2	0.250	60.40	70.55	78.54	62.67	54.11	21.13	24.67	24.05
2.5	0.177	87.35	91.78	97.04	83.67	69.67	61.87	67.33	77.82
2.75	0.149	98.43	98.11	97.75	91.56	87.83	87.65	94.05	95.76
3	0.125	98.55	98.33	98.25	94.67	94.76	98.65	97.43	98.67
4	0.062	98.70	98.78	99.24	98.43	98.67	98.68	97.65	98.86
5	0.031	99.11	99.17	99.57	99.44	99.43	98.78	99.01	99.07
6	0.015	99.24	99.64	99.84	99.76	99.65	99.46	99.54	99.15
7	0.007	99.92	99.95	99.93	99.92	99.96	99.87	99.76	99.54
8	0.003	100.00	100.00	100.00	100.00	100.00	100.00	100.00	100.00

Source: Estimated by the researcher (through field survey and laboratory work).

represented in Fig. no. 10.11. The curves are identical in shape suggesting similar hydrodynamic conditions involved during the period of sedimentation and deposition.

Table no. 10.5 reveals the fact that weight percentage of sediments at all the eight stations varies from the lower part of the basin to upper part. At the sample site A sand size is 41.86. Fine sand is found at sample site station D. Silt range between 3-5 and it is highest in sample site E (4.6). Likewise, weight percentage is also less than 5 are of clay i.e. 1.09. At all stations weight of coarse sand varies from 2.42-44.34. Fine sand from 56.89 to 96.53 weight percentage, Silt size from 0.13-4.6 and the percentage weight of clay ranges between 0.5-1.09.

Table - 10.4

GRAIN SIZE DISTRIBUTION OF BANK SEDIMENTS

Sample site	Mean size (mm)	Standard deviation	Kurtosis	Skewness	Shorting co-efficient
	m_2	D	KG	SK	SO
A	0.21	1.53	1.01	-0.71	2.12
B	0.30	0.97	1.03	-0.72	1.67
C	0.37	0.65	1.05	-0.74	1.33
D	0.35	0.94	0.86	-0.55	1.86
E	0.36	1.23	0.77	-0.40	2.95
F	0.17	0.48	0.98	-0.76	1.13
G	0.16	0.50	1.76	-0.67	1.07
H	0.24	0.46	1.11	-0.75	1.12

Source: Estimated by the researcher (through field survey and loboratory work).

Table - 10.5

TEXTURAL CHARACTERISTICS OF BED MATERIALS OF SAMPLING SITE A - H

Sample sites	Diameter (in mm)	Coarse sand (>1-0.50)	Fine Sand (0.50-0.125)	Silt3-5 (0.125-0.031)	Clay (<5< 0.031)
A	Weight percentage	41.86	56.89	0.37	0.88
B	Weight percentage	30.75	67.69	0.89	0.67

C	Weight percentage	16.81	80.96	1.34	0.89
D	Weight percentage	25.5	69.4	4.6	0.5
E	Weight percentage	44.34	51.63	3.41	0.62
F	Weight percentage	2.85	95.88	0.18	1.09
G	Weight percentage	8.74	88.7	1.6	0.96
H	Weight percentage	2.42	96.53	0.13	0.92

Source: Estimated by the researcher (through field survey and loboratory work).

BED SEDIMENT ANALYSIS: The size and composition of sediments along the longitudinal profile at the river quite varies. Bed sediment analysis is very important to know the size and composition of sediments as well as hydrologic characteristics of the stream. Grain size sediments determine the resistance of erosive agents along the river.

Analysis of the grain size distribution of bed materials and it has been shown by a set of histograms for the eight sampling sites (Table no. 10.6 and Fig. no.10.12).

Table and Figure portrays that at site H, the bed sediment of size 0.5 mm to 1.00 mm represents highest weight percentage of 65.03 and sediment size of 0.015 mm represents lowest weight percentage of 0.12. The mean size of sediment is 1.56 mm. The values of statistical parameters such as standard deviation, kurtosis, skewness and shorting coefficient (Table no. 10.7) reveals the fact that the sediment are well sorted medium sand but moderately peaked and negatively skewness. At the G the mean size of sediment is 1.56 mm which indicates that on an average the sediment is of coarse type. The sediment of size of 0.500 0.354 mm represents highest weight percentage of 32.61 and 0.003 mm to 0.007 mm size. The values of statistical parameter (Table no. 6.7) reveals well sorted coarse sediments with moderately peaked and negatively skewed distribution. It also portrays that the well shorted coarse sediment with moderately peaked and negatively skewed distribution. At all the site the distribution of sediment characterized by mean size value of 1.48 which is indicative of the coarse type of sediment size. The bed sediment size of 0.500 0.354 mm represents highest weight percentage. The sediment size of less than 0.031 mm represents only 0.06 weight percentage.

Table - 10.6

BED SEDIMENT SIZE DISTRIBUTION AT SITE A-H

Dia. In f	Dia. In mm	CUMULATIVE WEIGHT PERCENT							
		A	B	C	D	E	F	G	H
-1	2	4.60	13.25	0.35	1.85	2.31	1.84	0.89	5.76
0	1	4.12	17.89	0.38	5.65	3.73	4.52	2.44	14.74
1	0.500	10.45	34.65	2.53	34.43	19.07	30.83	28.54	65.03
1.5	0.354	41.75	41.70	32.75	55.64	43.15	57.43	63.34	84.37
2	0.250	64.40	50.85	67.87	69.72	70.54	75.72	80.84	92.84
2.5	0.177	91.85	77.65	76.74	92.75	94.43	94.54	97.54	97.63
2.75	0.149	94.43	96.64	97.04	97.75	98.26	98.32	97.87	98.85
3	0.125	98.65	99.74	98.54	98.28	98.65	99.17	99.26	99.04
4	0.062	98.71	99.88	98.74	98.90	98.77	99.28	99.56	99.34
5	0.031	99.48	99.93	98.87	99.33	99.59	99.65	99.65	99.54
6	0.015	99.64	99.94	99.44	99.58	99.87	99.79	99.87	99.68
7	0.007	99.82	99.98	99.63	99.89	99.94	99.94	99.93	99.80
8	0.003	100.00	100.00	100.00	100.00	100.00	100.00	100.00	100.00

Source: Estimated by the researcher (through field survey and loboratory work).

BANK SEDIMENT PROPERTIES (A)

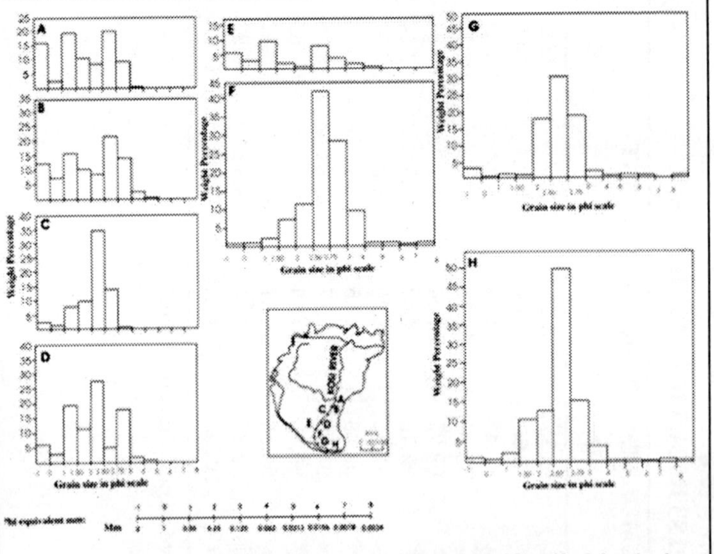

Histograms showing size distribution of bank sediments of Sapt Kosi river basin

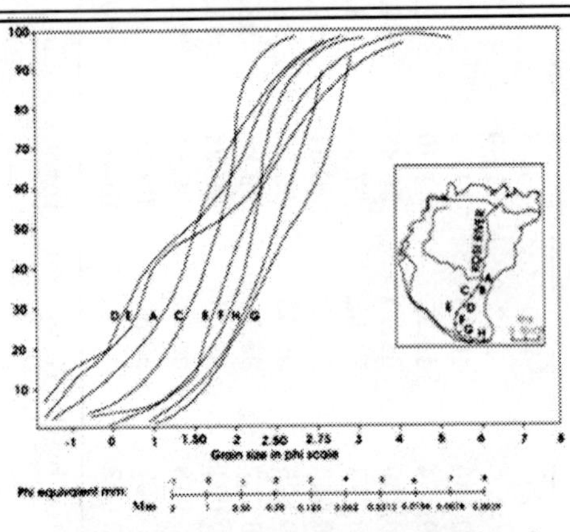

Cumulative curves showing size composition of bank sediments of the Sapt Kosi river basin

Fig. 10.11

The mean size of sediment is 1.48 mm indicating deposition of coarse sediment and weight percentage of 3.52 in the Terai region and on the river bed. The sample site is located on the upstream of the confluence point of all the tributaries and main flow of Sapt Kosi river. Sediment load of Sapt Kosi river basin based on the intensive field work and its measurement from the selected stations have been analyzed therein.

The value of statistical parameters of F site (Table no. 10.7) such as standard deviation, kurtosis, skewness and sorting co-efficient are 0.71, 0.98, -0.16 and 1.43 respectively. It indicates that the sediments are well shorted coarse sand and skewed towards right direction. At the D the size distribution has mean size of 1.40 mm indicating the presence of coarse sediments. The sediment size of 1.00, 0.50 represents highest weight percentage. It is indicative of shorted coarse sediment and negatively skewed. At the C site, sediment size varies with 0.354 0.520 mm. The mean size of sediment is 1.48 mm. The sample is located in the upstream of the confluence point and presence ricomparatively an extensive agricultural field in the valley area may be the cause of yielding coarser materials. The mean size of bed sediment in the stream upstream is 1.96 mm and in the extreme downstream is 0.88 mm which is indicative of surfacial erosion in the basin is the more active along the downstream section of the river.

Table - 10.7

GRAIN SIZE DISTRIBUTION OF BED SEDIMENTS

Sample Sites	Mean Size (mm)	Standard Deviation	Kurtosis	Skewness	Sorting Co-efficient
	m_2	D	KG	SK	SO
A	1.96	0.70	1.06	-0.27	1.32
B	1.42	1.25	0.70	-0.82	2.04
C	1.87	0.51	0.85	-1.64	1.21
D	1.40	0.86	0.87	-0.50	1.48
E	1.33	0.56	1.77	-0.16	1.35
F	1.45	0.71	0.98	-0.16	1.43
G	1.56	0.68	1.06	-0.23	1.37
H	0.88	0.76	0.91	-0.38	1.54

Source: Estimated by the researcher (through field survey and loboratory work).

Table - 10.8
TEXTURAL CHARACTERISTICS OF BED MATERIALS

Sample sites	Diameter (in mm)	Coarse sand (>1-0.50)	Fine Sand 1.3 (0.50-0.125)	Silt 3-5 (0.125-0.031)	Clay (<5< 0.031)
A	Weight percentage	10.51	88.06	0.87	0.56
B	Weight percentage	34.64	65.1	0.19	0.07
C	Weight percentage	3.52	95.05	0.24	1.19
D	Weight percentage	35.5	63.41	0.64	0.45
E	Weight percentage	19.08	79.62	1.24	0.06
F	Weight percentage	30.01	69.2	0.53	0.26
G	Weight percentage	29.13	70.33	0.16	0.38
H	Weight percentage	64.8	34.1	0.63	0.47

Source: Estimated by the researcher (through field survey and loboratory work).

SEDIEMNT LOAD: In a geomorpological study of a river basin sedimentological characteristic of a bed, bank and sandbar material is considered to be an important aspect, because it records the nature of sediment erosion, sediment load and deposition in different sections of the basin. It is also very helpful in the analysis of collected samples and helps us to know the history of sedimentation under varied fluvio-geohydrodynamics and sedimentation environment. Grain size is the most suitable parameter for reflecting the regional effect that the sediment erosion has to undergo. Size distribution of sediment grains and their shorting behavior particles clues to the condition of function of sediment and sediment load. The texture of sediment is usually described by statistical parameters such as mean size, standard deviation, the cumulative curves prepared from the idea of grain size analysis, kurtosis, skewness and shorting coefficient which intern reflect the environment of deposition, shorting characteristics, coarse and fine distribution of sediments and distance of transportation of sediments.

The movement of sediment by flowing water is a complex problem. Even in its simplest form, represented by flow in a channel of simple geometrical shape. It reveals a very high

BANK SEDIMENT PROPERTIES (B)

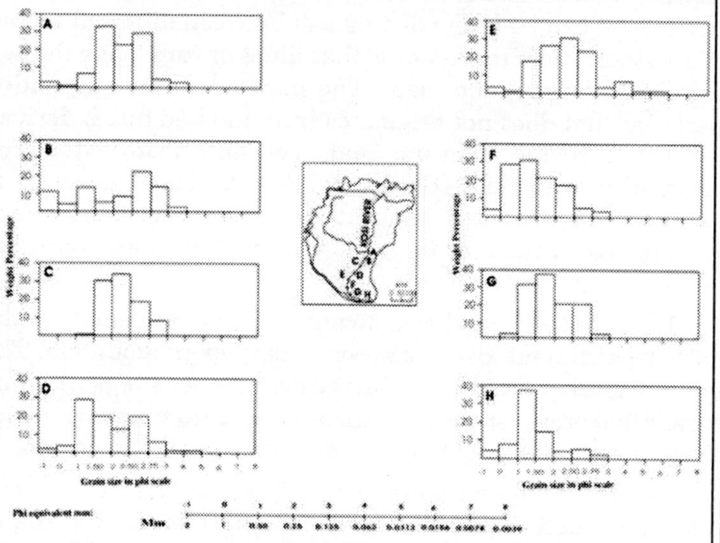

Histograms showing size distribution of bank sediments of the Sapt Kosi river basin

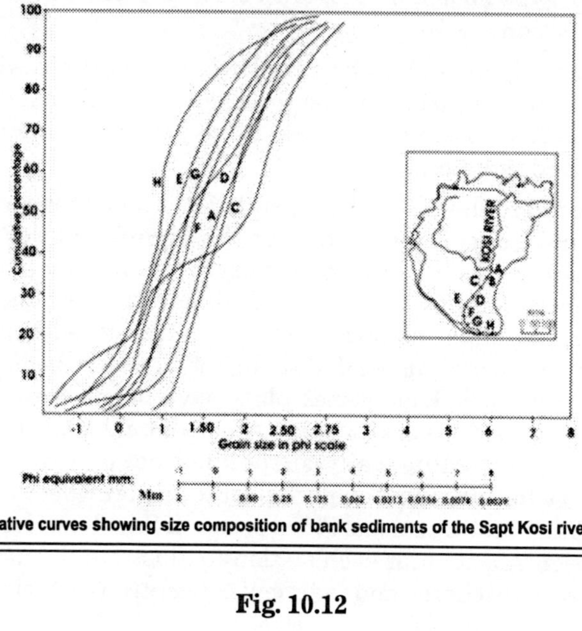

Cumulative curves showing size composition of bank sediments of the Sapt Kosi river basin

Fig. 10.12

degree of up steady non-uniform flow condition. Because the stream bed changes continuously and their by influences water, as well as material, flow. On the basis of movement of sediment particles it can be easily distinguish between different modes of transportation: the material that slides or rolls along the bed are called 'suspended load'. The portion of fine suspended particles that does not originates from the bed but is derived from the erosion from upstream bed and head water area referred as 'wash-load (Bruun, P., 1963).' Bed loaded transport as obtained a certain magnitude by which shear stress transferred by the flow to the bottom push the loose particles along.

In Sapt Kosi river basin owing to the heterogeneity of the surface, sediment load is also very complex in mountain. The general gradient is 61 m/km (in the mountainous region) which means that erosion power is tremendous due to stream gradient. Furthermore the gully being narrow, it is further increased.

Sediment components of any types constitute a) first, adetrital function (Pebbles sand and mud) brought to site to the deposition of some source area and b) Chemical function (Calcite, gypsum and others) form at or vary near to the site of accumulation.

The components may be mixed in any proportions, yielding sediment ranging from smaller to larger size, sediment derived from the river basin constitutes: i) Suspended sediment load and ii) Bed sediment.

Suspended sediment load: suspended bed sediment carries by flows and considered as one of the significant reason of geomorphic effectiveness. Sapt Kosi is a perennial river; but suspended sediment transport dominates in the monsoon season. Principal characteristics of suspended sediment concentration and suspended sediment load have been shown in Table no. 6.9. Time series plots have been given for the temporal variations in discharge and sediment concentration. The data regarding daily suspended concentration and discharge for the monsoon period for four sites where used for the sediment rating curves. Flood events of Sapt Kosi river basin have been selected for each station to illustrate the nature of changes, in discharge and sediment concentration during flood event.

The general characteristics of suspended sediment concentration and suspended sediment load in Sapt Kosi river basin exhibited in Table no. 6.9. The sediment concentration and discharge are very low in non monsoon season and measureable only during the monsoon season. The mean suspended sedimentation ranges between 1,145.41 ppm g/l in Arun river to 3,245.44 ppm g/l in Sun Kosi. In Sapt Kosi river the mean monthly sedimentation concentration is higher at Chatara-Kothu 3,537.75 ppm. The maximum suspended concentration ranges between 4,990.43 ppm g/l at Bagmati river and 11,821.4 ppm g/l at the lower tract of the Sun Kosi river. Tamur river possesses very high suspended sediment load (57.56 X 10^6 tons/year) where as in Arun river it is only 0.87 X 10^6 tons/year. Sediment yield found very high in Tamur river (10,205 ton/km^2/year) where as it is very low in Arun river (1,430 ton/km^2/year).

Sand Bar sediment characteristics: Sand bar are the important morphological features bearing on channel by hydraulics. The analysis of sand bar sediment helps to understand the placement of sediments and carrying capacity of the water in the river, compositional and textural characteristics of sand bar materials of Sapt Kosi river basin are analyzed and a systematic description of their characteristics is presented in the Table no. 6.10.

The sand bar sediments have been collected from four sample sites, that is B, E, G and H and the grain size analyze have been carried out and a set of four histograms have been drawn (Fig. no. 10.13) and Table no. 10.10.

A cumulative curve in order to compute the statistical parameters have also been prepared (Fig. no. 10.9). It portrays that the size distribution of sand bar materials from B-H predominated by the coarse sediments. It is most identitical with a mean size of 1.0 mm and 1.7 mm respectively.

The sediment class of 0.50 mm to 1.00 mm represents highest weight percentage of 48.82 where as the sediment size of 0.0313 mm-0.0156 mm represents lowest weight percentage of 0.048 at sample site H. at the sample site D the sediment size of 1.00-

Table - 10.9
SUSPENDED SEDIMENT CONCENTRATION AND SEDIMENT LOAD

PARAMETER	ARUN RIVER	BAGMATI RIVER	TAMUR RIVER	SUN KOSI
Basin Area* in km^2	34,660	13,589	6,741	3,857
Suspended sediment concentration (gram/liter)#				
Mean (ppm)	1855.47	1145.41	2711.45	3245.44
Minimum (ppm)	488.89	132.78	903.34	866.1
Maximum (ppm)	6904.46	4990.43	6948.2	11821.4
Suspended sediment load$				
Sediment load (10^6 tons/year)	0.87	-	57.56	-
Sediment yield (ton/km^2/year)	1430 (Tribeni)	Sampkhel 1,480 (Site: 550.1)	10,205 (Mulghat, Site: 690)	3 950 (Tribeni)

Source: * Calculated by researcher (through field survey and loboratory work)., # DHM, Nepal; $ International River center

0.50 mm represents the lowest weight percentage of 0.02 mm. the statistical parameter and textural classes represents the identical distribution and textural characteristics of the sediments. The sediment size of more than 0.50 mm represents weight percentage of 57.16 and 55.191. At sample site H and D the sediment size of 0.250-0.177 mm at site B represents the highest weight percentage and the site G the sediment size 1.00-0.50 mm represents highest weight percentage of 38.73. The statistical parameters of grain size reveals that the sediments area very well sorted negatively skewed and moderately peaked (Table no. 10.11).

Conclusively, it may be pointed out that the sampled sediments from the different fluvio- sedimentary environment in the basin which are analyzed and computed reveals and imaging facts. The grain size of sediment is generally found to decrease downstream with concomitant increasing in the length of river but the increase in grain size indicates anomalies. It's also indicating of coarse sediments from the surface runoff during flash floods and influx of coarse grains from tributaries having relatively higher transporting power.

FLOOD PLAIN MORPHOLOGY: Floodplains and their minor morphologic subdivisions are primarily depositional landforms. define a genetic floodplain as "the largely horizontally bedded alluvial landform adjacent to a channel, separated from the channel by banks, and built of sediment transported by the present flow-regime." This definition precisely emphasizes the linkage between active alluvium, floodplains, and the hydraulic geometry of modern rivers. Floodplains are built in two fundamental ways, by lateral accretion and by vertical accretion. As such, their forms are genetically related to specific depositional processes (Figure 10.14 & Table 10.3).

Table - 10.10

SANDBAR SEDIMENT SIZE DISTRIBUTION AT SITE B, E, G & H

Dia. In f	Dia. In mm	CUMULATIVE WEIGHT PERCENT			
		B	E	G	H
-1	2	1.95	7.48	17.16	15.49
0	1	2.88	16.35	27.54	20.78

1	0.500	4.76	65.17	68.44	55.81
1.5	0.354	8.00	79.74	82.37	64.09
2	0.250	20.73	89.29	88.40	76.34
2.5	0.177	64.66	97.72	98.27	93.67
2.75	0.149	95.74	98.38	98.43	98.74
3	0.125	99.24	98.53	99.57	98.59
4	0.0625	99.47	99.50	99.76	99.56
5	0.0313	99.83	99.73	99.84	99.63
6	0.0156	99.89	99.84	99.87	99.86
7	0.0078	99.97	99.98	99.97	99.89
8	0.0039	100.00	100.00	100.00	100.00

Source: Estimated by the researcher (through field survey and loboratory work).

Table - 10.11

GRAIN SIZE DISTRIBUTION OF SANDBAR SEDIMENTS

Sample Site	Mean size (mm)	Standard deviation	Kurtosis	Skewness	Shorting co-efficient
	m_2	D	KG	SK	SO
B	0.95	0.25	1.22	-0.75	1.07
E	0.93	0.52	0.93	-0.02	1.22
G	0.80	1.25	0.81	-0.26	1.31
H	1.03	1.04	0.94	-0.25	1.18

Source: Estimated by the researcher (through field survey and loboratory work).

Table - 10.12

TEXTURAL CHARACTERISTICS OF SANDBAR SEDIMENTS

Sample sites	Diameter	Coarse sand	Fine Sand 1.3	Silt 3-5	Clay
	(in mm)	(>1-0.50)	(0.50-0.125)	(0.125-0.031)	(<5<0.031)
B	Weight percentage	5.85	93.78	0.19	0.18
E	Weight percentage	54.12	45.31	0.1	0.47
G	Weight percentage	8.45	91.12	0.05	0.38
H	Weight percentage	55.94	42.95	0.86	0.25

Source: Estimated by the researcher (through field survey and loboratory work).

LATERAL ACCRETION: Point bars are the most important component of lateral accretion. As a meandering channel migrates across the floodplain, the cut bank on the convex side is undercut and eroded. In streams that carry relatively coarse alluvium and lack natural levees, lateral accretion may account for 80-90 % of the floodplain deposition.

VERTICAL ACCRETION: Large rivers with gentle gradients and a predominantly suspended load typically have meandering channels on broad floodplains. When they flood, the velocity in the overbank water may be very low, and the floodwaters may be weeks in receding into the channel. Under these conditions, floodplain deposition is largely by vertical accretion. The deposits are usually fine sand or mud although splays from breached channel banks may bring coarse sediment. Very large areas of the lower Sapt Kosi river floodplain were submerged for weeks during the great flood of 2008. Natural levees and ox bow lakes are common in this situation.

During the period 1807 to 1873, when the river was flowing direct into the Ganga the rate of the lateral movement was relatively slow. After that, since Sapt Kosi has begun to flow into the Ghughri, the rate of translation has markedly increased. The river was jacketed in the fifties and the tendency of shifting was considerable modified after that.

A better understanding of the sediment supply in the upper catchments of the rivers may be extremely rewarding in developing long term solutions to flood mitigation with afforestation and land use changes. This is where 'integrated management' of river basin becomes crucial (S., Rajiv, 2008,).

PAST HISTORY OF THE RIVER SAPT KOSI: From present topography, past reports and local sayings, it appears that Sapt Kosi formerly flowed Far East of its present course. More than a century and a half ago, Dr. B. Hamilton recorded the belief amongst the local people that the river, on reaching the plains, instead of running almost directly south to join the Ganga as it was about 50 years back, was flowing considerably eastward and joined the Ganges far downstream of its present confluence. He formed an impression that Sapt Kosi, during the remote past, possibly flowed south-east and then towards the east and joined

Table - 10.13
GENETIC CLASSIFICATIONS OF VALLEY SEDIMENTS

PLACE OF DEPOSITION	NAME	CHARACTERISTICS
Channel	Transitory channel deposits	Primarily bedload temporarily at rest; part may be preserved in more durable channel fills or lateral accretions
	Leg deposits	Segregations of larger of heavier particles, more persistent then transitory channel deposits, and including heavy mineral placers
	Channel fills	Accumulations in abandoned or aggrading channel segments; ranging from relatively coarse bedload to find-grained oxbow lake deposits
Channel Margin	Lateral accretion deposits	Lateral accretion deposits Point and marginal bars that may be preserved by channel shifting and added to overbank floodplain by vertical accretion deposits at top
Overbank floodplain	Vertical accretion	Fine-grained sediment deposited from suspended load of overbank flooded water, including natural levee and backland (backswamp) deposits
	deposits splays	Local accumulations of bedload materials, spread from channels onto adjacent floodplains
Valley margin	Colluvium	Deposits derived chiefly from unconcentrated slope wash and soil creep on adjacent valley sides
	Mass movement deposits	Earthflow, debris avalanche, and landslide deposits commonly intermix with marginal colluvium; mudflows usually follow channels but also spill overbank

Source: S.C. Happ, in Vanoni, 1971. Compare than below drawn

BANK SEDIMENT PROPERTIES (C)

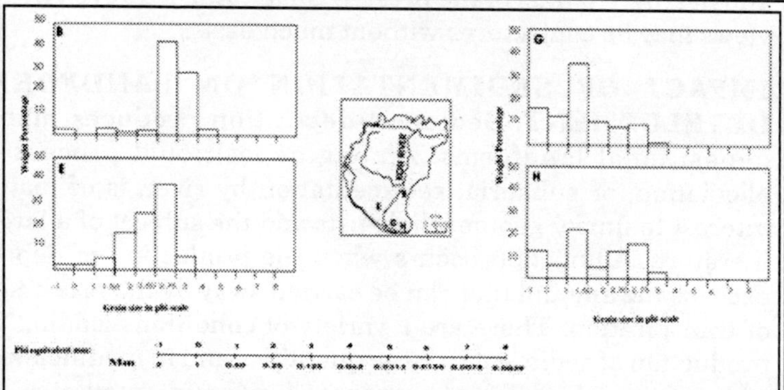

Histograms showing size distribution of sandbars and sediments of Sapt Kosi river basin

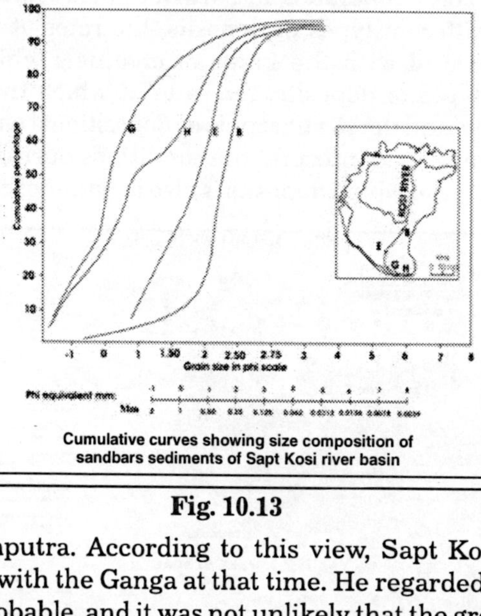

Cumulative curves showing size composition of sandbars sediments of Sapt Kosi river basin

Fig. 10.13

the Brahamputra. According to this view, Sapt Kosi had no connection with the Ganga at that time. He regarded this view as highly probable, and it was not unlikely that the great chains of lakes and marshes north and east of Malda (West Bengal) were the remnants of a great river bed formed by the United Kosi and the Mahananda. He also considered it probable that afterwards, when Sapt Kosi joined the Ganga, the united river opened up the wide channel now known as the Padma at Jangipur in West Bengal, through which the Ganges is still flowing. This aluvision caused the Ganga to desert and leave its original course below Jangipur, which is the comparatively dry and narrow channel of the present Bhagirathi. However, these views may be conjectures without much basis.

IMPACT OF SEDIMENTATION ON LANDFORM DEVELOPMENT: Sediment deposition produces many depositional landforms. Among depositional processes alleviation, or subaerial sedimentation by river, is of major interest to fluvial geomorphologists and the subject of a large literature. Alluviation occurs when the production of debris exceeds the amount that can be carried away by the processes of transpiration. There are a variety of conditions leading to production of sediment in large quantities and to modification of regimens of sediment transport large and small stream channels. These conditions and with the kinds of channels which produce different types of deposits, the rates at which these conditions and with the kinds of channels which produce different types of deposits, the rates at which these deposits accumulate, and the stratigraphic or depositional characteristics that can be used to interpret the conditions during deposition. Size of material which deposits is also a matter of concern such

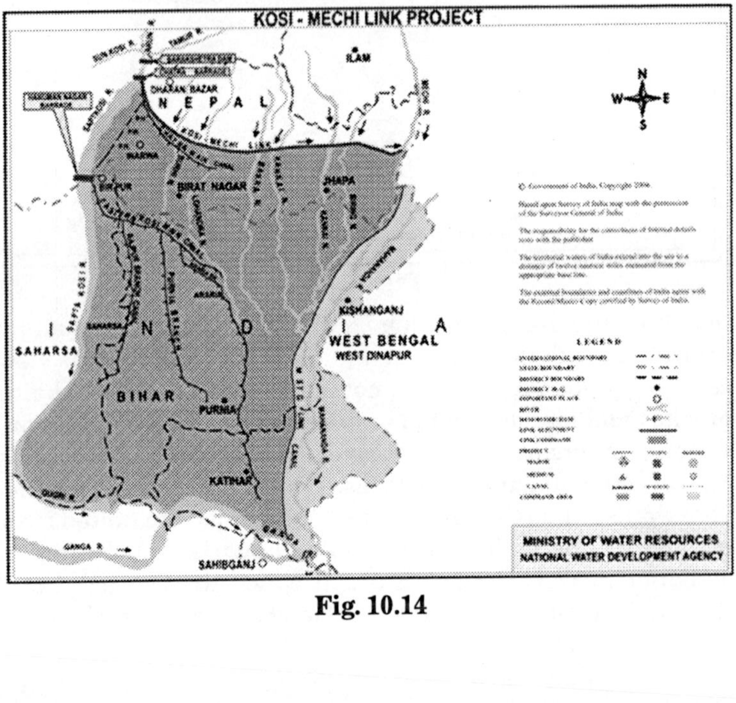

Fig. 10.14

as in the upper basin there is large size (boulders) are lying on the channel bed, in middle course of the river there are medium size deposition of boulders, pebbles and gravels (specially in Terai region), in upper Sapt Kosi mega fan area there is large

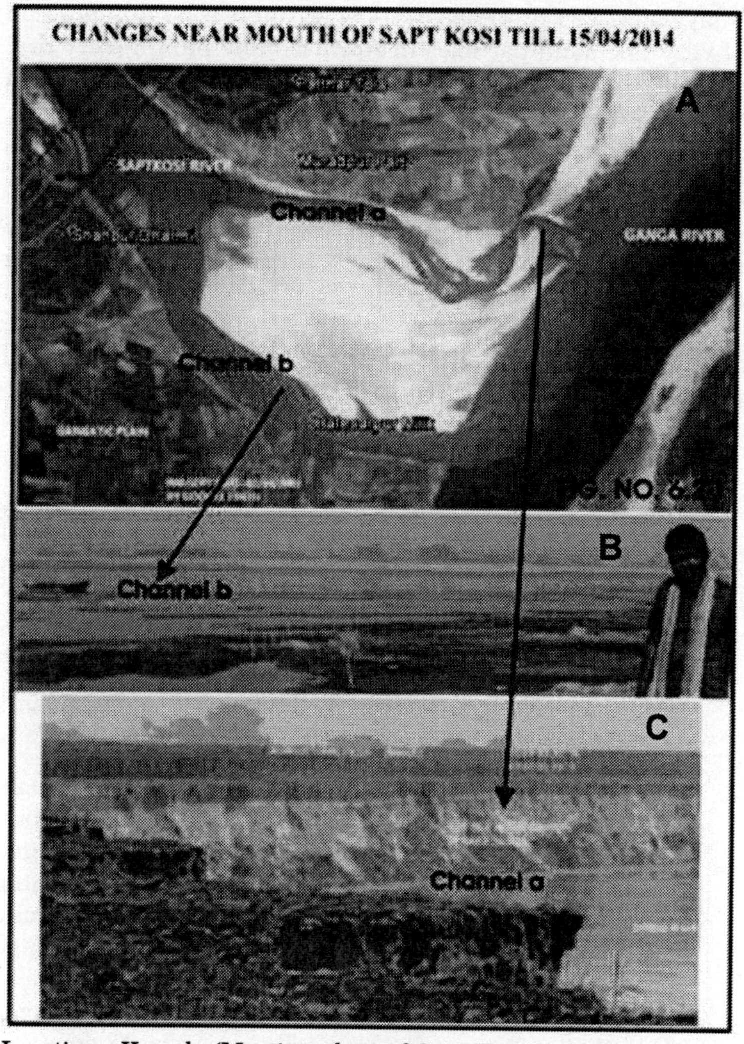

Location : Kursela (Meeting place of Sapt Kosi and Ganga River), Bihar, India.

Fig. 10.15

distribution of big size of sand particles where as at the last course area near Kursela there are fine to very fine sand particles are distributed. These patterns of deposition directly correlated with formation of landform of entire basin (created by deposition process). Examples as in north of Naugachia, Vijayghat (25^{0}25'40" N and 87^{0}04'47" E) there is large deposition of very fine sand particles in April to June month (very dry) creates Aeolian landscape as Sand dunes of different pattern which is very migrating landscape in character.

RECENT CHANGING BEHAVIOR: We know that this kind of highly discharging and sedimented bed surface are highly changeable. These changes are traceable with compare of one man's general life span. So any one can trace the changing behavior of the river. In respect of Sapt Kosi river basin several changes occurred in the plain tract of Bihar where sedimentation and shifting of channels are simentanously playing. Some of them our observations are follows:

CHANGES NEAR MOUTH OF SAPTKOSI TILL 15/04/2014: the following changes identified by researcher during her field visit (Fig. no. 10.15):

1. We are viewing two separate channels (namely channel '**a**' and channel '**b**' in above Figure but when we went for ground checkup than there was only single channel which was channel 'b' where as channel 'a' was vanished.

2. Middle bar or sand deposition was totally vanished, and mostly submerged by water mass.

3. In the Google map above we can clearly see that in the left embankment of Ganga possesses deposition of sand bar but during our field excursion I seen there was Katni (bank erosion) active; local civilians loss their some Katta of land through this Katni and it is still in active mode. The main reason for this katni is Sapt Kosi river's water mass which contains very fine sediment (sand particles) and it is responsible for bank erosion in left hand side of cliff Ganga. The inflow sediments of Sapt Kosi river is clearly visible in Fig. no. 10.15 (a).

CHAPTER – 11

BANK EROSION

Water flowing down to the sea over and immediately beneath the land surface, is the dominant agent of landscape alteration. Near-surface weathering and groundwater solution provide a load for flowing streams, and mass wasting may dump great quantities of rock debris at the foot of slopes, but eventually rivers must carry all but a small fraction of the total rock waste from the lands to the sea. Estimates of the relative importance of various agents to continental denudation can hardly be better than orders of magnitude, but one careful evaluation (Garrels and Mackenzie, 1971). Rivers and streams are products of their catchments. They are often referred to as dynamic systems which mean they are in a constant state of change. River bank erosion is a natural geomorphic process which occurs in all channels as adjustments of channel size and shape are made to convey the discharge and sediment supplied from the stream catchment. However, increases in sediment supply due to accelerated stream bank erosion are often linked to land-use change and are a major contributor to sediment yield in rivers.

CAUSES OF BANK EROSION: The factors controlling river and stream formation are complex and interrelated. These factors include the amount and rate of supply of water and sediment into stream systems, catchment geology, and the type and extent of vegetation in the catchment. As these factors change over time, river systems respond by altering their shape, form and/or location. In stable streams the rate of these changes is generally slow and imperceptible.

STREAM BANKS ERODE FOR MANY REASONS: Stream bank erosion is a natural process that over time has resulted in the formation of the productive floodplains and alluvial terraces common to the middle and lower reaches of many of India river systems. Paradoxically, even stable river systems have some eroding banks. However, the rate at which erosion is occurring in stable systems is generally much slower and of a smaller scale than that which occurs in unstable systems. Events like flooding can trigger dramatic and sudden changes in rivers and streams. However, land use and stream management can also trigger erosion responses. The responses can be complex, often resulting in accelerated rates of erosion and sometimes affecting stability for decades. Over-clearing of catchment and stream bank vegetation, poorly managed sand and gravel extraction, and stream straightening works are examples of management practices which result in accelerated rates of bank erosion. Erosion can also be accelerated by factors such as:

i. Stream bed lowering or infill
ii. Inundation of bank soils followed by rapid drops in flow after flooding
iii. Saturation of banks from off-stream sources
iv. Redirection and acceleration of flow around infrastructure, obstructions, debris or vegetation within the stream channel
v. Removal or disturbance of protective vegetation from stream banks as a result of trees falling from banks or through poorly managed stock grazing, clearing or fire
vi. Bank soil characteristics such as poor drainage or seams of readily erodible material within the bank profile
vii. Wave action generated by wind or boat wash;
viii. Excessive or inappropriate sand and gravel extraction
ix. Intense rainfall events (e.g. cyclones).

Sediment gets into the river not only from the catchment area but is also contributed by erosion of its banks. Cause of failure of river banks can be any of the following singly or in combination.

i. Underwater erosion along the toe of bank during the falling stage of the river.
ii. Direct erosion of the river bank.

iii. Sloughing of saturated bank caused by rapid drawdown.

iv. Liquefaction of saturated silty and sandy bank material.

v. Erosion due to seepage from banks at low river discharge.

vi. Scour along waterline due to wind or wave wash of passing vessels.

Failures under 1, 3, 4 and 5 depend more on the properties of the bank material than direct erosive action of stream flow.

PROCESSES OF STREAM BANK EROSION: The various mechanisms of stream bank erosion generally fall into two main groups, bank scour and mass failure. In many cases of bank instability both will be evident, often with either scour or mass failure being dominant. Bank scour is the direct removal of bank materials by the physical action of flowing water and is often dominant in smaller streams and the upper reaches of larger streams and rivers. Mass failure, which includes bank collapse and slumping, is where large chunks of bank material become unstable and topple into the stream or river in single events. Mass failure is often dominant in the lower reaches of large streams and often occurs in association with scouring of the lower banks. By looking carefully at the processes operating at a site it may be possible to narrow down the probable causes of instability. Correct diagnosis of the underlying causes of erosion problems is important if successful and cost-effective solutions are to be generated. Assistance with identifying the causes of bank erosion and advice on potential solutions is available from the Department of Natural Resources and Water.

BANK SCOUR: Bank scour is the direct removal of bank materials by the physical action of flowing water and the sediment that it carries. Undercutting of the bank toe is a sign of bank scour As flow speed increases, the erosive power of flowing water also increases and scour may occur. Increases in flow speed can be the result of natural (e.g. from the previous page d, f and i) and/or human induced processes (e.g. a and e). Undercutting of the bank toe is an obvious sign of scour processes. Effective strategies for combating scour are generally aimed at reducing flow speed through revegetation and in some cases through strategic bank or channel works.

MASS FAILURE: Mass failure describes the various mechanisms of bank erosion that result in sections of the bank sliding or toppling into the stream. Mass failure is sometimes described as collapse or slumping. Bare and near-vertical banks or areas of slumped bank materials are obvious signs of these processes. The causes of these types of failures are often difficult to determine but can include natural (e.g. b or f) and/or human factors (e.g. c, e, g or h). Collapse following undermining of the bank toe and slumping as a result of saturation after flooding are common examples of mass failure. Effective strategies for combating slumping or bank collapse are generally aimed at stabilizing the bank toe and restoring bank vegetation.

BANK EROSION INCLUDES TWO MAIN GROUPS OF PROCESSES:

a. Hydraulic processes at or below the water surface entrain sediment and directly contribute to erosion, particularly of non-cohesive banks, by processes of bank undercutting, bed degradation, and basal cleanout.

b. Gravitational mass failure processes (including shallow and rotational slides, slab and cantilever failures, earthflows and dry granular flows) detach sediment primarily from cohesive banks and make it available for fluvial transport.

BANK EROSION TYPES: There are several types of bank erosion acting in the Kosi river basin, main two types of force acting on bank erosion that is gravitational failure mechanisms and another is hydrocally induced failure mechanisms, which are described as follows.

GRAVITATIONAL FAILURE MECHANISMS:

SHALLOW SLIDES: These are shallow failures where a layer of material moves along a plane parallel to the bank surface. They are typical of soils with low cohesion, and occur when the angle of the bank exceeds the angle of internal friction of the bank material. Typically they occur where banks are moderately steep. This types of bank erosion generally founds in middle and upper reaches of the basin where the general gradient is much more.

ROTATIONAL FAILURES: i) Rotational failures are deepseated movements of material both downward and outward along a curved slip surface and are common on cohesive banks with slopes less than 60°. ii) Commonly a result of scour at the base of the bank and/or high pore water pressure within the bank material. iii) Often they will occur during rapid drawdown following high flow events. This types of bank erosion is generally seen in Terai region where subterranean water is dominant as underground streaming.

SLAB FAILURE: Slab failures are the sliding and forward toppling of a deepseated mass into the channel. They are associated with steep, low height, fine-grained cohesive banks and tend to occur during lower flow conditions. They result from the combination of scour at the bank toe, high pore water pressure in the bank material and the development of tension cracks at the top of the bank. This type of bank erosion is greatly found in upper and middle Himalaya of the basin where steep gradient is found. River water cut their lowering materials and overloaded materials finaly slab down due to slide on a slide plane. The recent Bhote Kosi disaster of bank erosion was the result of slab failure.

CANTILEVER FAILURE: Cantilever failures are the collapse of an overhanging block into the channel, often occurring after the bank has been undercut. They tend to occur in a composite of fine/coarse grained materials, and to be active during low flow conditions. This types of bank erosion is greatly active in Sapt Kosi fan area where meander forms and cuts their bed due to cantilever failure.

WET EARTHFLOW: This type of failure occurs where the loss of strength of a section of bank due tosaturation increases bank weight and decreases bank material strength to a point where the soil flows as a highly viscous liquid. Failures typically occur on low angle banks and the affected material flows down the bank to form lobes of material at the toe. Wet earthflow is active in the areas of wetted river bed. In the middle Himalayan region of the basin possesses high amount of monsoonal rainfall in rainy season, therefore the bank erosion across the channel got active due to wetted channel side beds. Due to this landslide accelerate in many region in the baisn.

POPOUT FAILURE: Popout failure is a term used to describe failures where small to medium sized blocks are forced out at or near the base of the river bank due to excessive porewater pressure and overburden. A slab of material in the lower half of the bank will fall out, leaving an alcove-shaped cavity. This mode of failure is usually associated with steep banks and saturated finer-grained cohesive bank materials that allow the buildup of positive pore-water pressure and/or strong seepage within its

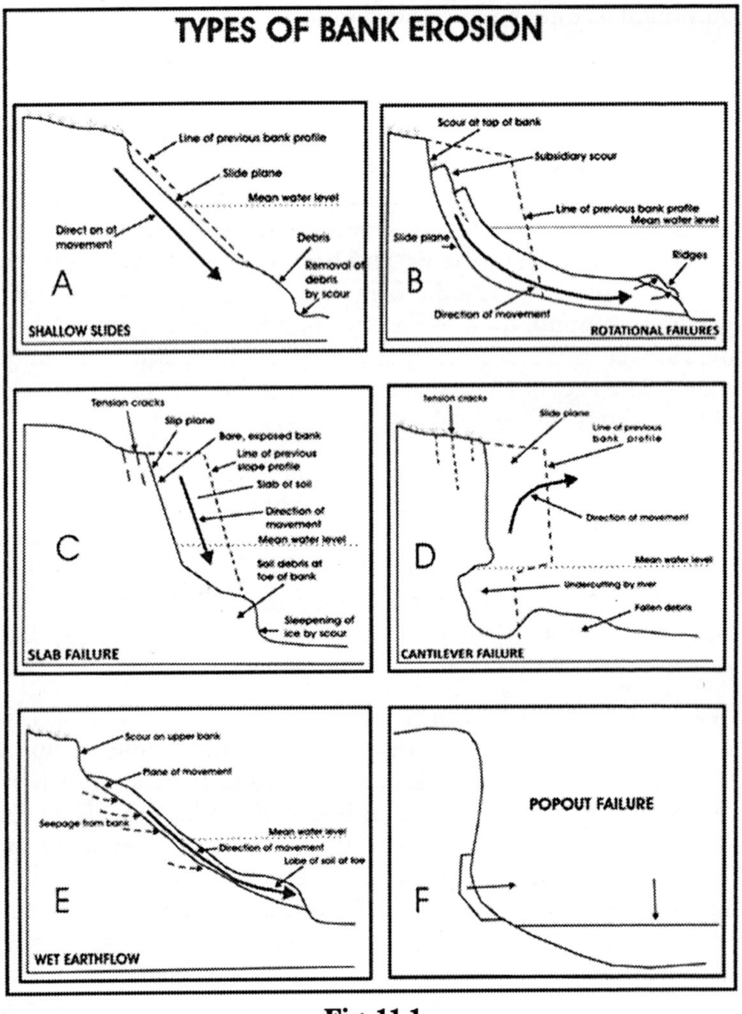

Fig. 11.1

a stream undercuts the toe of sand, gravel, or deeply weathered rock bank. Again this type of bank erosion is massively found in Terai and Sapt Kosi mega fan area.

PIPING FAILURE: This is the collapse of part of the bank due to high groundwater seepage pressures and rates of f low causing selective removal of sections of the bank. It is usually due to preferential ground water flow along interbedded saturated layers contained within stratified river banks, with lenses of sand and coarser material sandwiched between layers of finer cohesive material. This type of bank erosion occours in the plain region of all Sapt Kosi mega fan area of Bihar plain.

HYDRAULICALLY INDUCED FAILURE MECHANISMS (FLUVIAL EROSION): Direct fluvial erosion results from the change in balance between hydraulic shear stress and bank material strength. Where the shear stress exceeds bank material strength sediment transport will be initiated. This occurs because shear stress increases as flow increases, while bank material strength typically reduces (e.g., when the bank becomes saturated). There are three main types of hydraulically induced failures. Fluvial erosion is often a precursor to gravitational failures, and is also responsible for transport of the debris produced by gravitational failures. Hydrologcially Sapt Kosi river basin possesses in strange amount of force due to high gradient of landmass in the upper catchment area of the basin. When the river comes in Terai their hydrolocally mechanisms comes with great force. Boulders, pebbles and cobbles come and deposits here. But with the induce of water mass any type of huge erosional works occurs, particularly during maximum fluvial seasons. 2008 flood was a good lesson of hydrologically induced failure mechanism which was occoured in Sapt Kosi east embankment near Kushaha in Nepal.

FACTORS AFFECTING STREAM BANK EROSION: The stability of a bank primarily depends on channel and flow characteristics, and the strength of the bank materials (Table below). I) Instability can be inherent in some channel systems as a result of the nature of the river system (e.g., high energy braided rivers) and historic or geomorphic factors (e.g., tectonic uplift).

Table - 11.1

FACTORS AFFECTING STREAM BANK EROSION

FACTOR	RELATIVE CHARACTERISTICS
Stream frequency	Rainfall intensity and duration
Flow properties	Magnitude-frequency, duration and variability of stream dischargeMagnitude and distribution of stream velocity and shear stressDegree of turbulenceSediment load
Bank material	Size, gradation, cohensively and composition stratification of bank material
Bank geometry	Height, slope, length, profile shape
Bank moisture conditions	Soil moisture levels, seepage, pore water pressures, piping
Channel geometry	Width, depth, slope of channel, stream curvature (concave, convex, straight)
Vegetation	Type % cover, age, rooting depth, exposed roots, stability
Man-induced factors	Stock and vehicle usage, artificial drainage input

CHANNEL CHARACTERISTICS AND DYNAMICS: Lateral and vertical movement of thalweg at different locations in an alluvial river from year to year is described both in qualitative and quantitative terms from the study of river cross sections. The lateral/ vertical movements of the river bed, where possible, are co-related with the problem of bank erosion/collapse, if any. The description of secondary channels, especially those frequently shifting and causing significant changes in flow pattern near the banks are described. There are several channel characteristics through which we can understand the fluvial dynamics of Sapt Kosi river basin, such as bank full width, regime status, river bed characteristics, sediment transport, flow characteristics, plan form, bars and shoals, bed forms, bed and energy slopes and channel migration etc.

BANK FULL WIDTH: Bank full width is a subjective term. This is taken as the river width at water surface level

Table - 11.2
WIDTH-DEPTH COMPUTAION

S.N.	River	Station	Zero Level	Danger Level	Depth	Bankfull Area	Bankfull HMD	Width (AIHD)	Width Depth
			(m)	(m)	(m)	(m)	(m)	(m)	(w/d)
01	Baghmati	Dheng	65.00	70.10	5.10	500	2.25	222.22	44
		Hayaghat	35.00	45.72	10.72	775	6.00	129.17	12
02	Kamala-Balan	Jaynagar	64.00	67.75	3.75	120	1.04	115.38	30
		Jhanjharpur	44.33	50.00	5.67	450	4.00	112.50	20
03	Kosi	Birpur	-	-	-	-	-	-	-
		Baltara	25.75	33.85	8.10	3000	4.75	631.58	78

Source: CWC, Govt. of India, 2014.

corresponding to the dominant discharge or bank full discharge. This could also be decided by the visual examination of the river cross sections and confirmed with the study of gauge and discharge curve plotted on semi-log paper (the level at which the curve flattens out could be taken as the bank level).

Mean Depth (D) = Area of cross section at bank full stage (A) / Bank full width (B) Width Depth Ratio = B/D. It is also useful to study changes in channel parameters in the downstream direction for different frequencies of flow.

REGIME STATUS: From the regime formulae of Lacey, width, depth, area and velocity are worked out and these are compared with the observed values at different river stages. Such a study would indicate the applicability or otherwise of the regime formulae of Lacey.

RIVER BED CHARACTERISTICS: It is obvious that soon after stream channels begin to incise, they should encounter the relatively unweatered rock beneath the soil and regolith. 'Many mountain streams have their channels entirely in rock, with only scattered patches of alluvium and colluviums on the channel floor and walls. Even in hilly regions of subdued relief and deep soil cover, rock outcrops are common in stream beds (Bloom, 1978).' The qualitative and quantitative descriptions of lateral slope of the flood plain are given from the observed cross sections for different years.

Aggradation and degradation of the river bed in a reach are studied through a comparative study of river cross-sections for different years. These crosssections for different years are superimposed and the area of each cross-section below a reference line is worked out. An increase in the area from one year to another would indicate degradation while a decrease in area would indicate aggradation at the site. 11 Aggradation and degradation of the bed may also be reflected in a shift in the G-D curve over a given period. For a meaningful study, it would be desirable to draw G-D curves separately for rising and falling stages of the river. The above study from cross-sections and G-D curve would indicate aggradation/ degradation at a particular cross-section and not in a reach. However, such a trend between two sites could be broadly studied through a sediment balance study for the reach.

SEDIMENTS TRANSPORT: Though there are numerous sediment transport formulae developed by various investigators from time to time, none of these is considered suitable for all situations. Efforts are, therefore, made to develop relationships between observed values of sediment transport on one hand and different flow parameters on the other. The different flow parameters could be discharge, velocity and stream power.

FLOW CHARACTERISTICS: Relations could be developed by plotting graphs on log – log scale between width, depth and velocity on one hand and discharge on the other. For low flows, critical velocity may be co-related with depth to develop an equation which will have the form of Kennedy's equation.

Table - 11.3
COMPARATIVE STREAM FLOW VOLUME

RIVER BASINS	DUDH KOSI	TAMA KOSI	LIKKHU	TAMOR
	4918	4153	2091	9510
4-6 km	2130	829	248	2194
Glacier Melt	869	253	81	723

Unit in Million Cubic Meters, mcm

Source: D. Alford and R. Armstrong, 2010, p. 494.

PLAN FORM: Shape of river in plan is very important in many design problems concerning location of bridges etc. In general, the plan forms of alluvial rivers can be classified as (i) Braided, (ii) Straight and (iii) Meandering. Meandering and braiding reaches of the river are separately studied in chapter no. two and four.

Meander length (also called wavelength) Vs. Bank full width, Meander width Vs. Bank full width, Radius of curvature Vs. Bank full width.

Plan Form index represents the percentage of actual flow width per braided channel. Obviously, this index reflects the fluvial landform disposition with respect to a given water level and its lower value is indicative of higher degree of braiding.

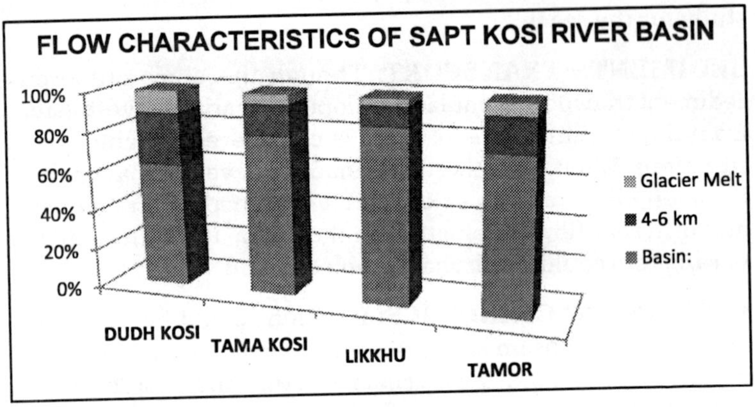

Fig. 11.2

Flow Geometry Index reflects the underwater sub-channel disposition and the hydraulic efficiency of a braided stream. Its higher value signifies occurrence of higher degree of braiding. Cross slope is basically a form ratio indicator and its higher value indicates higher braiding intensity.

BARS AND SHOALS: Bars and shoals in the river bed may be identified as point bars, middle bars, alternate bars, islands, etc. 'Singh et al. (Singh, Parkash and Gohain, 1993) documented different bedforms/bars and their internal structures within the braided and straight reaches of the Kosi River flowing at present through the western margin of the megafan. They also provided important information on the stratigraphy of the Kosi megafan from shallow and deep borehole data collected from the modern river beds as well as from other areas of the megafan.' The changing features of bars/islands associated with meanders/ braiding, etc. may be ascertained from the sequential satellite images of different resolutions as per requirement to identify the probable direction of flow pattern/shifting of river courses for consideration in correcting river alignments. Temporal Satellite data may be used to study the river dynamics and area statistics to be derived from the temporal data to find the erosion and deposition characteristics. Unwanted shoals are identified for removal.

CHAPTER – 12

APPLIED GEOMORPHOLOGY

'Applied Geomorphology may be defined as the application of geomorphic understanding to the analysis and solution of problems concerning land occupancy, resource exploitation, environmental management and planning (Jones, D.K.C., 1980). 'Geomorphologists tend to consider all geomorphological research to be "applied" in the sense that each advance in knowledge provides a clear view of how the earth works, we must all be applied scientists (Crag and Craft, 1982).' 'The scope of applied geomorphology is basically related to the application of geomorphological knowledge and understanding for economic development and well being of the human beings. It is done precisely to understand the complex nature of interaction between human activities and geomorphological processes and agents (Singh & Ojha, 1996). Thus the main theme of applied geomorphology is to investigate the impact and changes brought in natural environmental systems and geomorphic/environmental processes on different aspects of human society and human activities.

Landforms are the most common features encountered by anyone engaged in geomorphic field work. The purpose of this chapter is to point up specifically some of the applications of geomorphology to the types of problems that are commonly encountered by geologists, geographers and some engineers.

H. Th. Verstappen, 1983, has outlined the following areas of interest where geomorphic knowledge and techniques may be applied for understanding and solving problems:

1. Topographic and thematic mapping of natural resource.
2. Survey of natural hazards, landslides, avalanches, earthquakes, volcanism, land subsidence, flooding and

droughts and natural hazard reduction.

3. Rural development and planning emphasizing land utilization control of man-induced soil erosion, channel manipulation and river basin (watershed) management.

4. Urbanization, mining and construction.

5. Engineering design.

In recent years application of remote sensing and Geographical Information System (GIS) played multidimensional aspect of Applied Geomorphology. Approximate all type of applied aspect can be directly real time monitor and plan. Data accuracy (work perfection) is almost 100% true. These spatial technological tools give freedom to perform Geomorphologists on optimum level with good accuracy level. Through application of these techniques goemorphologists are able to plane and make strategies of resource consumption and mitigation for human welfare in a

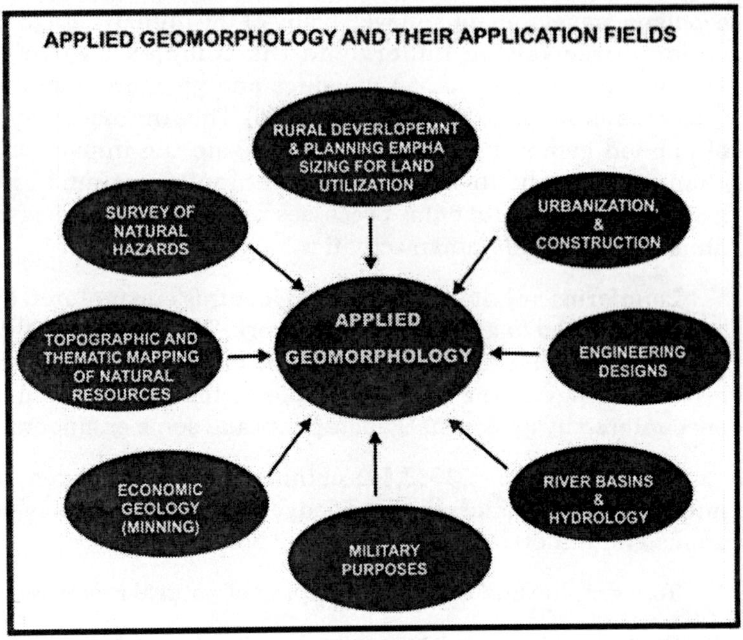

Fig. 12.1

sustainable way.

The science of geomorphology has important applications in the fundamental activities of man. It can serve as the basic source of planning in a wide range of fields, such as economic, geology, hydrology, land-use, military science and also in the field of diverse engineering projects. For a systematic approach application of the knowledge of geomorphology can be classified as under:

i) Application of geomorphology to Land Use.

ii) Application of geomorphology to Hydrology.

iii) Application of geomorphology to Economic Geology.

iv) Application of geomorphology to Military purposes.

v) Application of geomorphology to Engineering projects.

APPLICATION OF GEOMORPHOLOGY TO LAND USE: The application of geomorphology can effectively be done in planning land-utilization. Chatterjee emphasizes the application of geomorphology, it can render in the development and rational utilization of natural resources (Chatterjee, 1965). By land utilization is here meant the augmentation of various agricultural, industrial and cultural activities that have land as the primary factor of setting. Here overall emphasis in planning land-utilisation, as Kaushic pointed out, is made to consider such aspects like agriculture, horticulture, pastoralism, industry, transport and human settlement, rural and urban (Kaushic, 1967). Land is a very important natural resource that needs to be effectively managed to make the optimum use of this valuable resource. Though, land use is controlled by several factors like soil, hydrologic, climatic, socio-economic and political, yet geology and geomorphology play an important role. The basic landscape formation depends mainly on the inherent lithology and structure and it is the landscape that directly affects development of certain landuse/landcover and inhibits other. Information on existing landuse/ landcover, its spatial distribution and change are essential pre-requisite for planning (Dhinwa, Pathak, Sastry, Rao, Majumdar, Chotani, Singh, Sinha, 1992). Therefore, the application of geomorphology can be apply for the benefit of the people of Sapt Kosi river basin.

AGRICULTURE: Augmentation of agricultural production is the most basic need. 'It forms the major landuse class and occupies the central portion of the study area which is the valley depression and filled by fertile alluvium, where the Siwaliks have been masked by Dun Gravels. The second level of classification shows fallow land, plantation and standing crop in this region. Fallow land has also been identified by its lighter tone due to the absence of vegetation and moisture and regular pattern whereas crop land is identified by pinkish tone (Rehman, RAO and Saran, 2013).' In the Indian part of the river basin is highly and densely populated in the another hand flood and sand deposition (sedimentation in the upper Sapt Kosi mega fan) are common problem for destructing thousands of hectare cultivable fields and in the highly lofty mountainous region extensively lies high degree of slope which again is not much suitable for commercial agriculture or other economic activities. It is geomorphological applied aspect through proper planning and management would be possible to convert the land in to productive.

The augmentation of agricultural production can be done by paying consideration to the following points:

i) Preparation of land-use maps.

ii) Estimating ground and sub-soil moisture during different crop season.

iii) Estimating the extent of soil and land erosion and supplying land conservation measures.

iv) Examining per-acre yield in the different types of land surface and re-organisating crop pattern and cultivation according to geomorphology.

From the relief point of view there are five broad categories of agricultural land.

SOUTHERN MEGA FAN AREA: This agricultural region is highly fertile due to deposition of very fine sediments carrying out by Sapt Kosi river's per year flood phenomenon. But due to flood in monsoon season there is few agricultural activities are happening. But the high nutrition value of the soil it compensate farmer by winter ravi crops as very high yielding.

These plots of Sapt Kosi river basin, having bestowed with climatic conditions of the monsoon type, rainfall being more than 200 cm. annually is suitable for paddy, sugarcane and jute cultivation. Geomorphic principles applied in this regard are:

Thus the region is a highly saturated land surface; water management may play a dominant role in the distribution of river water as channeling system. The irrigation water supply should be such that crop evaporation is not hindered by a shortage of water. Percolating soil moisture conveys soluble salts by advection downward which leaches the root zone but may create at the same time adverse effects by pushing the water table upward in the winter where as in monsoon water should be discharge by better management.

NORTHERN MEGA FAN AREA: This includes the northern portion of Bihar state of India, particularly Supaul, Madhubani, Saharasa, Sheohar, Darbhanga, and Madehepuara districts and some part of Nepal also. Due to larger size of sediments sandy soil are largely found. It should be manage by soil fertilization. Forestation may be helpful to achieve this goal. Jute or Makhana may be the best option for planting in monsoon season.

THE TERAI: Thus the Terai region is marked by deposition of big size of boulders and pebbles under deep thickness. Due to this a thin thickness of soil is found. Deforestation is common in this region. Thus some farmers are successfully growing paddy and some vegetables through forming organic farming. Thin sandy soil may be helpful for improving fertility.

HIMALAYAN REGION: High degrees of slopes are main reason for development of commercial farms. Therefore, the sloping land should be converted into plots of step size farm one above another. In the same time, contour plaguing may be helpful to protect the life of stepped plots against soil erosion, particularly through sheet or rill erosion. It is also an important measure for preserving the soil productivity. Water can effectively be stored in individuals plots for a required period without damaging the adjacent plots.

Table - 12.1
IMPLEMENTATION OF GEOMORPHIC IDEAS IN AGRICULTURE

S.N.	GEOMORPHIC REGIONS	AGRICULTURAL PROBLEM	GEOMORPHIC IMPLEMENTATION	PREDICTED RESULTS
01	Southern mega fan area	Flood and high silt	Increasing the number of channelization from both side of the river	Water will be divide and reduce the impact of flood level
02	Northern mega fan area	High sedimentation of sand size materials	Afforestation and water management	Sandy soil will be develop
03	The Terai	High rate of deposition of boulders and pebbles	Afforestation and organic farming	Average yield may be possible
04	Himalaya region	High degree of slope, soil erosion, and communication	Contour plaguing, small and step size farms, small channel for water passing	Soil erosion will be reduce and soil fertility may sustain for longer time
05	Tibet Plateau	Unavailability of soil resource, water scarcity in summer, about no vegetation cover	Store water in a small chackdams, organic farming	Due to water storage some small surface vegetation will grow

Source: Table created by researcher, (through field survey).

MOISTUREFUL RIVER VALLEY OF TIBET PLATEAU: In the far Tibet plateau and different tributaries of Arun river in the upper catchment area, there is lack of deep soil resource. But in the river valley where moisture is available they cultivates. Maize and some vegetable with paddy are common but some instance to fruit can be also produce. Water resource are bitterly using by Chinese company as packed drinking water supply to other part of China.

HORTICULTURE: The sloping surface along the middle and Shivalik Himalayan region providing well drained ground can be converted into fruit gardens and other plantation gardens by applying proper calcium and organic fertilizers. In this part of the region fruits like papaya, bananas, jack-fruits and citrus fruits can give a good yield and can be grown on commercial basis.

Rainfall in the region being high (about 200 cm annually), temperature is mild in the middle Himalaya, the geomorphic setting of the region is such that there is common presence of uniformly sloping surfaces having average depth of coarse of mountainous soil. These uniformly sloping surfaces can affectively be converted into tea-gardens, which have ample scope in this part of the region.

PASTORALISM: on the Terai and middle Himalayan region where agricultural activities are difficult, animal husbandry is good option. Green sub tropical tree leaves are favorable for animals. Medium degree of slope in Terai region up to 600 meter where grassland are extensively available, pastoralism can be beneficial in providing ample scope for dairy-farming. Such as Hen, Goat and pig husbandry are more suitable in the region.

INDUSTRY: Application of geomorphology to the future industrial activities in the area may be examined as regards:

a. Availability of raw materials
b. Availability of space
c. Adequate supply of water
d. Availability of abundant land.

In and around the region of Sapt Kosi river basin, there is scarcity of raw materials specialy for mineral resource. But water, soil, and forest are extensively available resources. In the middle and upper reaches of the basin water resource is mostly found. There is huge amount of availability of water resource through which hydrological electricity can be made. Most of upper reaches of Arun river small hydro power plant is going to built up. Certain clay for specialized purpose, such as potteries, fire-bricks are available. Other raw materials like timber are available in this part of abundance in middle and greater Himalaya. Industries based on the products of forests which themselves are a gift of terrain and geomorphology can be established.

The region can be proper site for certain manufacturing industries, such as timber-industries, small hydro electricity plant, and other various such small industries which require cheap availability of land.

TRANSPORT: Geomorphic application to transport is to be examined in the setting of transport lines, types and availability of raw materials. Basement of the region that provides generally sloppy, materials like boulders in Terai region and in channels of different rivers, stone-chips are available in channel bed again. In the upper reaches of Himalaya and Tibet there is special need of geomorphologist who can assess the risk of natural hazard such as landslide, soil creep, glacier lake outburst and avalanches. Over the high degree of slope in the greater Himalayan region building of transport system is tough one but a geomorphologist can build the tough road. In the plain region there is densely network of railway and roadway. A new Mahasetu (Longg bridge) has been built near Nirmali with more than 56 pillars and it is part of East-West corridor (N.H.). It is second time when bridge has been constructed previously, British bridge were blown away by the river.

In the periglacier region of Tibet there is special need of geomorphologist, because road are not permanent due to periglacier processes. Thus high quality of concrete using for constructing, and mostly along the river valley. During monsoon season the changes of landslide increases, so geomorphologists can identify such region of troubling.

HUMAN SETTLEMENT: Application of geomorphology to

human settlement – rural and urban may be examined from the point of view of their dwellings, the product of the environment. Rural house-types like in any other part of Sapt Kosi river basin have been adjusted to the physical environment. Geomorphology is a major part of this. Application of geomorphorphology in constructing dwellings is seen in the selecation of site which generally forms the high degree of slopes in the middle and higher Himalaya creates so many problems. Selection of such kind of dwelling site should be choose where low gradient is available on the surface or where minimum the risk of landslide is predictable in the Himalayan region. In the flood region of mega fan area there should be keep in mind the dwelling house should be comparatively on higher place where minimum chances for flood water reach. In the Terai region channel of any stream should be avoid because in rainy season these channels may be dangersous for flash floods. So beside the river (where elevation is available) and above the levees should good place for settlement establishment.

Building material required for construction of such dwellings are earth (soil), sand, boulders, wood-rafters and bamboos which are locally available. Walls are earthen-built, both ground and walls are earthen-plastered, roof is covered by earthen-tiles particularly in the plain region of the Sapt Kosi mega fan area. Urban settlements always give rise to departure from the physical or geomorphic setting. That is why in the region of Bihar urban centers like Madhubani, Madhepra, Saharasa, Darbhanga, and Supaul there is water logging problem particularly in the monsoon season because drainage system is not developed according to geomorphological slope of the surface. Any such similarity is maintained at the cost of neglecting geomorphic or physical conditions of the region which is variable throughout the world. Generally in Tibet region settlements are found near and beside river channel because of during dry season it only the means of source for all purposes. Kathmandu valley one of the best example of settlement in the lacustrine valley. One time it was surrounded by a lake but due to vanishing of lake (due to some tectonic event) almost flat surface released and now it is one of the most densely settled grounds in Sapt Kosi river basin; but again here drainage contour should be follow by Kathmandu Municipal

Corporation through proper geomorphic planning of sewage system. Comparatively in the Terai region of the basin is densely populated than middle and higher Himalayan region, because there is availability of almost low degree of slope surface where all type of geomorphic planning could be possible. Local river basin is using as providing building materials such as sand, gravels, boulders and chips. In the Terai cities such as Siraha, Janakpur, Dhankuta and Chatra, etc. has densely populated.

Rural dwellings have imprints both of geomorphic setting on the one hand and modern urbanization on the other.

By applying geomorphic criteria human habitations can be placed in a better setting. These geomorphic conditions are that dwelling should be always constructed on the save place, save from sudden water logging in plain region or sudden water currenting in Terai region. There should be not use heavy building materials in the highly sloppy surface in Himalayan terrain, because there may be chances of landsliding in rainy season. Again light material such as bamboos, earthen tiles on the roof may be helpful to reducing the risk of landslide. Hence, through this construction cost can be curtailed. Applying these measures, human habitations can be placed in a better position with minimum cost.

APPLICATION OF GEOMORPHOLOGY TO HYDROLOGY: 'Hydrology problems are closely related to geomorphology of a particular terrain. These problems may result from high degree of gradient of the channel, presence of fractures, joints or faults (Prakash, 1978).' 'Joint intergovernmental policies on water resources and hazard mitigation should be established. Flood water may be turned into water resources if there are sufficient water storage facilities (Chen, Hu, Sh, Deng, Khanal, Zhu, and Han, 2013).' There are different types of hazards such as cloud burst, Land sliding, Flood etc. which are directly or indirectly related with surface and subsurface water bodies. Through the implementation of geomorphic ideas the risk and effect of hydrological problems can be reduce.

PERMEABILITY VERSUS IMPERMEABILITY OF TERRAIN:

PERMEABILITY: Permeability of the pediments surface along the ridges of the region is of course in a position to create hydrologic problems. It is because the surface in this part of the region is formed of coarse detritus eroded from the ridges and spread in form of 'fans'. These coarse detritus includes boulders, stone-chips (ranging between 1 cm to 5 cm), pebble and coarse quartz particles like sand etc., over such pediment surfaces, that originates at the foot of the ridges varies from a few meters in cross section to as much as a kilometer or more. Water received in these parts either from a direct rainfall or through 1st, 2nd or 3rd order channels that originate over the ridges get penetrated. As such the surfaces remain barren, devoid of any vegetation cover. Water-table over these parts is much deep. Hence construction of dam or tank would prove a complete failure, because of capacity of dam to store certain volume of water will not be attainable. Most of the water stored will be penetrated to underground through afforestation particularly in Shivalik and Terai region of the basin.

IMPERMEABILITY: on the contrary the site of a dam if selected on the Terai region crossing or along 4th, 5th or 6th order stream will be a scientific selection since the surface over these parts will be impervious. There will be the presence of numerous springs even in dry season too. The pattern is very similar to artesian type bedding plane. As such the application of geomorphology in selecting proper site cannot be overlooked.

OBTAINING WATER FROM WELLS: obtaining water from wells of large yield in permeable terrain, if not possible, is of course a difficult one especially in western part of Sapt Kosi mega fan area. On the other hand well construction on the lower parts to get large yield of water for irrigation purposes will prove most beneficial, since subsurface layer over these part is pervious in the entire Kosi fan region. Water spouts up after digging a few meters in depth in lower portion of Terai region because of high saturated layer of subsurface water.

GROUND WATER SUPPLY: Recognition of the possibilities of large supply of ground-water from glaciated areas largely depends upon familiarity with the type of deposits from which large yield can be obtained, along with an understanding of the geomorphic history of the area during periglacial times. Outwash plains, valley terrains and gravels are particularly likely to yield large volume of water. This ground water may be the major source of irrigation and use in other purposes in the plain tract of the basin.

SURFACE WATER SUPPLY: A large quantity of Pleistocene glacier is found around the greater Himalayan peaks in the basin which is actually the source of continuous supply of water during summer season and dry season of winter. That is why Sapt Kosi river basin supply huge amount of water all round the year. In the plain region near Kosi barrage east and west canal were constructed for supply surface water around the region. A number of projects also operated in upper and lower region of Sapt Kosi fan area, but still there is opportunity to develop many canal systems from west to east. Paleaochannels of the Sapt Kosi which is easily demark by satellite images can be use as regular supply of water during dry season. The best way to reduce the risk of flood is also micro level distribution of water resource from a single river system, this also the theory of Dutch water resource management system in Europe. In this system water will be flow freely in the all basin without creating hazard. Through this system water-table along these region will be observe near the surface which will be easily exploit through irrigation in agricultural sector too.

APPLICATION OF GEOMORPHOLOGY TO ECONOMIC GEOLOGY: An understanding of geomorphic features and history of a region may be profitably utilized in the exploration of certain mineral resource. Geomorphic features may serve in three ways, as guides in the search for minerals.

a) Topographic expressions of some minerals.

b) Topography of an area giving clue to geological structures that favour accumulation of minerals.

c) Geomorphic history of an area that appreciates conditions of mineral accumulations.

Thus, Sapt Kosi river basin possesses scanty of mineral resource, very less quantity of economic mineral resource is found in this basin. In the whole plain area is full of alluvial deposition which is mainly quaternary deposits suitable for agricultural activities. Some very few quantity of lignite is found in foothills of Shivalik but that lignite have very less economic value. Without this sedimentary rock as limestone and slate are extensively found in the upper reaches. But due to bed communication system there is less scope of economic exploitation of these minerals.

APPLICATION OF GEOMORPHOLOGY TO MILITARY PURPOSES: Importance of the application of geomorphology was only recognized since the World War II, because the effectiveness of a blitz depends to a large degree upon the trafficability of the terrain. As a consequence in the recent years increasing fervor is seen for terrain analysis or its appreciation for military purposes. A goemorphologists in this regard can be of best help, because his concept about terrain is far more adequate than those of the military specialists or other geologists for that matter. He appreciate that landforms are the result of an interaction of geomorphic and geologic processes through time. Landform are not the groups of unrelated and haphazard individual forms, but have systematic relationships that reveal their origin and tell much about the underlying bedrock, geology and structure, as well as the soils and vegetation of a region.

BASIS OF STUDY: Geomorphology can serve country's defense forces by way of terrain evaluation for different kinds of military purposes, air bases etc (Kaushic, 1967). For terrain evaluation the patterns, facets and sub-facets of the terrain are being noted by:

a. Air-photos interpretation and comparison.
b. study of topographic maps and lands
c. charting the lithology
d. analysis the soils
e. Studying the meteorological and climatic conditions.
f. Study of the flora and fauna.
g. Study of the socio-economic and cultural features.
h. satellite image interpretation and attributes classification
i. Mapping with GIS environment.

THE TYPE OF INFORMATION AN ARMY REQUIRES: Lt. Col. Kapoor (Kapoor, 1967), while paying attention towards the application of geomorphology in meeting some of the defense requirements, expresses that availability of terrain data is the most pressing problem hat a troop in war confronts. The information an army requires about inaccessible areas are:

a. Traffic ablility conditions prevailing during different seasons.
b. Suitability and problem of roads.
c. Airfield and camp construction with easy and rapidly.
d. Water-supply potential of the area.
e. Availability of accessibility of resources.
f. Concealment and camouflage the area provides.

ROLE OF AERIAL PHOTOGRAPHY: Aerial photos provide a technique and a tool to compare two non-contiguous areas with a high degree of accuracy. In inferring conditions of inaccessible areas, it is therefore vital that all our evaluation is based on the aerial photos and their interpretation. It is in the interpretation of these aerial photos that the two patterns is established, their facets can also be located the air-photos with a fair degree of accuracy and prediction be made achieving 60% to 70% accuracy.

To understand the geomorphic application of terrain, or identification and units, the Kosi river basin, and its environment may be examined.

TERRAIN UNITS BASED ON MORPHOLOGY OR SURFACE CONFIGURATION: Main landforms of the region, as outlined in chapter II, region has been divided into six macro units or facets.

1. The Plain, 2. The Terai, 3. The Shivalik, 4. The middle Himalaya, 5. The Great Himalaya, and 6. The Tibet plateau.

These are further divided into 21 Sub-facets in all are called micro-landforms units. Here these units can be expressed into condensed form.

WATER REGIME: On the basis of water regime of the area the undulating surfaces of Sapt Kosi river basin may further divided into two elements:

a) The permeable surfaces of Plain and Terai region. The water regime of these surfaces due to saturation of surface water network is very close to earth surface. But during dry season there is need to extra infiltration by channelization irrigation system.

b) The highland portion of basin towards north including Himalayan mountain range and Tibet plateau. In the Himalayan portion there is availability of monsoon rainfall in July-August-September month. Through this water is available throughout the year because of underground water infiltration.

It is thus found here that there are twenty terrain units morphologically and two according to water regime. All these have important geomorphic applications to defense forces, as regards trafficability conditions, suitability and problem of roads, airfield and camp construction with easy and quick, water-supply potential of the area, availability of accessible resources and camouflage the area may provide.

APPLICATION OF GEOMORPHOLOGY TO ENGINEERING PROBLEMS: Evaluation of geologic factors, terrain characteristics, geomorphic history of the area, surficial material, configuration of the bedrock profile etc. are important aspects of geomorphological application. Geomorphological techniques and knowledge become helpful in several engineering projects related to development activities e.g. construction of roads, dams and reservoirs (related to river valley projects), canals laying of pipelines for petroleum and potable water, withdrawal of groundwater etc. Some of the engineering projects may be examined. Certain engineering projects to be examined here are:

a. Highway construction
b. Location of sand and gravel pits
c. Domestic selection
d. Airfield site selection
e. Other applications

HIGHWAY CONSTRUCITON: The feasibility of highway construction largely depends upon the topography. Varying terrain unites, impose varying engineering problems or the varying terrains, there may be a karsts plain, glacial terrain, terrain of considerable relief, etc. In Sapt Kosi river basin the communication system well developed in plain region of Bihar

TABLE - 12.2
TERRAIN UNITS BASED ON MORPHOLOGY OR SURFACE CONFIGURATION

S.N.	MACRO UNITS	MICRO LANDFORMS	REMARKS
01	The Plain	Southern mega fan Kosi region	Very fine sediments
		Northern mega fan Kosi region	Sandy soil
		Eastern flooded region	Every year flood in rainy season
		Western uplands	Western boundary of the basin
02	The Terai	Terai	No flood water reach
		Bhabhar	
		Khadar	Sometimes flood water
03	The Shivalik	Churia mountains	Laterite deposits
		Mahabharat mountain	East west extension
		Upper shivalik	
04	The middle Himalaya	Eastern middle Himalaya	Provides hilly slopes for cultivation.
		Central middle Himalaya	
		Western middle Himalay	
05	The Great Himalaya	Polar and glacier landforms	Mt. Everest, Mt. Makalu etc.
		North of Great Himalaya	End of glaciers and moraine starts
		South of Great Himalaya	End of glaciers and feed to mostly southward river system
06	The Tibet plateau	Far west plateau	Less population
		Far east plateau	Less population
		Arun river valley	Some population
		Far north plateau	

Source: Based on Researcher's field survey and terrain analysis.

and Terai region where as in mountainous region road or highway construction is a tough work. High degrees of slope in those areas are obstacle for the construction. It has uplifted diastrophically during the Tertiary period as a side effect to the Himalayan orogeny. As a result the drainage network of the region began to cut their valley-beds deeply forming the flat-topped ridges, hills and flat-bottom valleys of considerable relief.

These areas of considerable relief like the region pose, certain geomorphic problems in the construction of highways, viz., require extensive cuttings and fillings, increased number of bridges, making over all costs of construction high. But the region has also certain favorable geomorphic aspects through which geomorphic ideas may be implement. The high mountainous region of middle and greater Himalaya possesses the problems related with landslides, earthquake and earth or mudflows.

A Sapt Kosi Mahasetu (E-W N.H. Corridor) across Sapt Kosi river near Nirmali in Bihar, India. This bridge has been construct in the place of former bridge which were blown away in 1956's flood.

It is directly correlated to topographic features of the concerned area mainly surface configuration. In the high mountainous region of Himalaya there is very tough task to construct roadway or railway system for communication. Local people in Higher and Middle Himalayas are used to cut the slope for pathway purpose. Along with slope (in river valley and gorge) a tiny 4-6 feet width pathway track construct which is very dangerous to walk over, because there cannot be possible to make a plain width road cause of high landslide zone and almost vertical slope. So, here local people use their applied geomorphological knowledge for communication purpose. But in Terai due to almost flat surface Nepal Government constructed National Highway extending from west to east with the help of Govt. of India. But due to high range of conglomerate deposition in the region it is very tough to stabilize the road for long time.

LOCATION OF SAND AND GRAVEL PITS: Sand and gravel have many engineering as well as commercial and industrial uses. These have distinctive topographic relationships, expressions, varying inherent qualities and possibilities of development. Some far north region of Arun river in upper reaches has sedimentary structure along the valley through which geomorphic knowledge for road construction may be implement there. Availability of building materials and energy are available in the river bed. Gravel is a chief source in the construction of all weather motorable roads. But in the areas of high angle of slope construction of road and other network is sometimes not possible because of landslide. Most of the roads of the border between Nepal and China have been constructed out of gravel (Unmettled road).

DAM SITE SELECTION: Application of geomorphology in the selection of dam sites including a synthesis of knowledge regarding terrains, their lithological and geological structure. Bryan (1929) points out at least five requirements of good reservoir sites:

a. A water tight basin of adequate size.

b. A narrow outlet of the basin with a foundation that will permit economical construction of dam

c. Opportunity to build an adequate and safe apill-way to carry surplus water.

d. Assurance of the life of reservoir against excessive deposition of mud and silt.

Generally, it supposes that Dam construction is the only solution for a problematic river basin such as Sapt Kosi. Since British govt. it was always a regional problem in the Sapt Kosi mega fan area, where almost every year flood commences. So many geologist and geomorphologists suggests to make a river valley project and some oppose it. These can be well understood in the following ways:

In the region under study Sapt Kosi river basin, high mountainous and lofty range are presence. These regions are areas of high risk of earthquake and landslides that is why the dam construction may accelerate the risk of these natural hazards. But small hydro power may be the solution for multipurpose objectives. These mountainous regions are highly favorable for generation of hydro power. Construction of barrage or dam is not possible in plain tract of Bihar region because of high sedimentation which is currently facing by Kosi barrage near Hanumannagar. Certain suitable sites for construction of dams are pointed out:

1. Across river valley of 3rd and 4th order stream of Tamur river near Dhankuta in Nepal
2. Across 5th and 6th order stream of Bhote Kosi river in Tatopani, Nepal.
3. Across 5th and 6th order of Dudhi Kosi, Nepal.
4. Across 6th order of Tama Kosi, Nepal.
5. Across 5th order of Arun river, Nepal

Besides, there are also suitable sites across many tributaries of 5th and 6th order in numerous areas or middle Himalaya. All these sites fulfill most of the requirements of good dam sites, e.g.,

a. These sites are excellent in respect of adequate volume of water,
b. Construction of streams so that streams across these parts form bottle-neck pattern.
c. Building materials are locally available so that it will make the construction of dam economical.

Dam sites should be avoided on permeable terrain, loosened sedimentary rock strata terrain, and landslide prone region too, otherwise it may prove a failure in the long lasting.

EMBANKMENT CONSTRUCTION: 'The popular flood control strategy in this region has been to construct embankments on both sides of the rivers which does provide some relief for a short time but over a long period, they have proved to be disastrous - one because siltation of river bed leads to a consequent rise in water level, and two, severe breaches in the embankment can cause huge loss of life and property of unwary people living along the 'secured' flood plain. As the embankments create a false sense of security among the people, they suffer the most, being completely unprepared to combat a flood (Sinha, 2008).' Therefore, there are so many controversies over construction of embankment along the river. Central and state of India and Bihar invested billion of money over construction of embankment along the Sapt Kosi ant its major tributaries but the problem never stopped. In some cases Sinha's statement about embankment construction proofed right way. So what is the future planning for this river. We should continue to construct embankment or live free to the river for flowing independently in the basin and let the river working in natural way without the influence of man.

AIRFIELD SITE SELECTION: Geomorphology has important role to play in the selection of site for airfield. A local climate of a region is of significance, but these too are related to geomorphic conditions to considerable degrees. The selection of airfield from the view point of civil purposes is to be considered taking into account.

a. Possibilities of long runways from several direction,

b. Drainage conditions

c. Amount of grading necessary

d. Dangers of flood and fog.

e. Water-supply

The need of long runways from several directions is a difficult problem to get fulfilled in an around the mountainous region where as it is tough to select such kind of runway in the in plain region because of flood. Over the different terrain of the region sites are suitable only for light to medium type of aircraft. But these sites have their strategic importance as regards concealment and camouflage the area provides. As regards drainage conditions the area is well-drained. But the great frequency of drainage lines is in itself a problem in making graded surface an important requirement. As regards floods, there is so danger the region would pose in plain tract, fog is also certainly dangerous casually in winter season. Southern Tibet plateau has some specific locations for airfield, but due to lack of population commercial flight may be in doubt, where as for Chinese military purpose it may be good option. On the whole the region has certain good spots of strategic importance for the construction of airfield. Those sites are:

Table - 12.3
SOME IMPORTANT SITES FOR AIRFIELD CONSTRUCTION

S.N.	NAME OF AIRFIELD SITES	COUNTRY	LOCAITON	REMARKS
01	MYANGLUNG	NEPAL	87°32'E/27°06'N	Flat top hill
02	DHARAN	NEPAL	87°17'E/26°50'N	-
03	CUO MUZHELIN LAKE	CHINA	88°06'E/28°24'N	N-E of the basin
04	DHALKEBAR	NEPAL	85°57'E/26°57'N	Piedmont (foothill) area of Shivalik
05	SONBARSA	INDIA	85°35'E/26°51'N	Near Nepal border

Source: Based on Researcher's field survey and terrain analysis.

Of these second and third are important strategically. These sites are at an altitude between 1,000 m to 2,000 m above the m.s.l. These provide suitable concealment and camouflage. But these sites may create problem, as regards construction of approach roads. The first one is important for making Chinese airbase camp for military purpose. Sakya (4280 m) is important for nearly road link to Cuo Muzhelin Lake. It is situated at a Tibet plateau flat leveled surface that provides radial drainage pattern sarrounding. Water-supply is also adequate here.

OTHER APPLICATIOINS: Application of geomorphology may be of practical use in tackling problem of soil erosion. The degree of slope of different landforms is an important factor in determining the processes of sheet wash, mass wasting, gullying and stream erosion. But it has also been observed that slope in itself is not always a controlling factor. Intensity of soil erosion also depends on the infiltration capacity of soils. It is found that areas of permeable soils having high degree of slope angles are highly susceptible to erosion. In the high undulating mountain of Himalaya it is common phenomena. These interrelated facts have frequently been observed in and around Sapt Kosi river basin except plain tract of Kosi. After taking into consideration these geomorphic facts, the problem of soil erosion can be tackled more scientifically.

Geomorphic principles may also have important application in classifying land types, which entail an appreciation of varying types of terrains and the best uses that may be made from the view point of agricultural productivity.

An example of GIS implementation in Flood management is given in Fig. no. 12.3. In Fig. no. 12.3 is stands for the final assessment values for flood risk covering the entire studies region where risk of flood is asserted as low, moderate, high and very high categories. Through this flood risk analysis map Engineers, planners, or Geomorphologists can easily plan to the area for recovering or planning purposes.

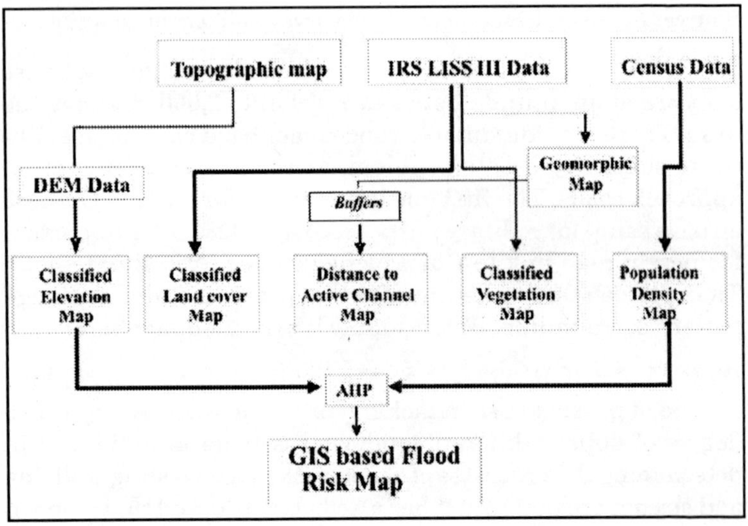

Fig. 12.2

Source: Bihar Govt. After R. Sinha, G.V. Bapalu L.K. Singh, and B. Rath

Besides their application in landuse planning, civil and hydrological engineering, soil surveying and conservation, they can be applied gainfully in various other fields, as indicated below (Nag & Saha, 1996).

SAPT KOSI PROBLEM AND BARRAGE CONSTRUCTION: So it will be seen that basic problem of Sapt Kosi is the silt, which produces spill which in its turn produces enormous flooding. With the modern techniques of dam construction control of spill from peak discharges has become a fairly simple affair but control of heavy amount of silt which is the crux of Sapt Kosi problem, has yet puzzled the engineer and unless that is solved Sapt Kosi problem cannot be said to have been completely solved. But, then we could not leave the things as it is. Some solution, if not entirely permanent, has to be found to stop this recurring menace of the river. If not for the many long years to come, something had to be done for the immediate future. Mr. Leopoled and Moddock, American experts who opined on Sapt Kosi problem, have rightly said that this appears to be one of the many instances in which this generation cannot

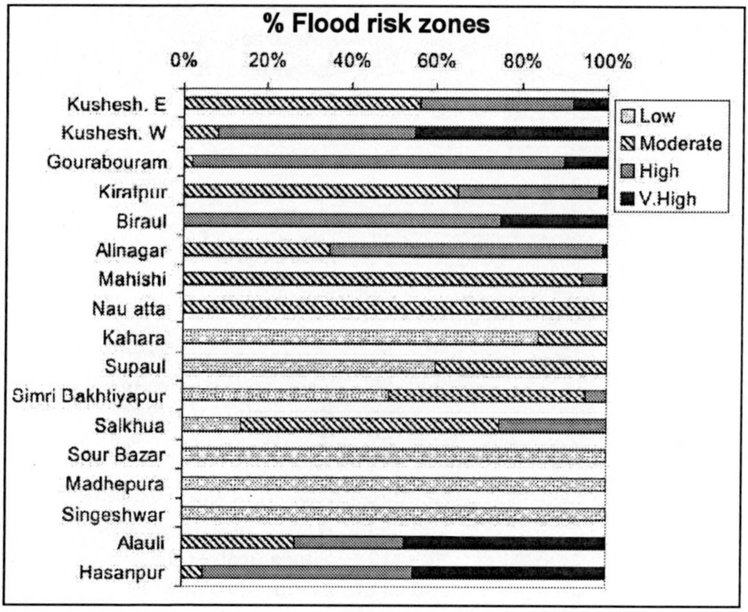

Fig. 12.3

solve the problems of the next generation and its own as well. as a matter of fact no major control scheme anywhere in the world offers complete flood protection or is a permanent solution to a flood problem. Sapt Kosi floods ae unusually complicated and any control method must be considered, not only impermanent, but also incomplete in that it offers a fair measure of protection and not a complete elimination of floods or flood damage.

Table - 12.4
APPLICATION OF GEOMORPHOLOGY

S.N.	Application Major Type	S.N.	Application at Micro Level
01	LANDUSE	01	Territory planning
		02	Regional area planning
		03	Conservation of the natural and cultural landscape
02	AGRICULTURE AND FORESTRY	01	Potential utilization
		02	Soil conservation
		03	Soil erosion control
		04	Reclamation of destroyed or new areas
		05	Soil reclamation
03	UNDERGROUND AND SURFACE CIVIL ENGINEERING	01	Reconstruction and replanning of settlements, especially of towns
		02	Designing of industrial buildings
		03	Communications (roads, railways, canals, and harbours)
		04	Hydro-engineering
		05	Reservoirs and dams
		06	Regulation of rivers
		07	Natural and artificial waterways
		08	Irrigation canals
		09	Harbor construction
		10	Shore protection
		11	Fishing projects
04	PROSPECTING AND		

EXPLOITING OF MINERAL RESOURCES	01	Prospecting
	02	Geological survey
	03	Exploitation
	04	Mining
	05	Potential and actual damage done by mining
	06	Reclamation of abandoned open-cast mines
	07	Landslip areas and regions of subsidence due to mining
	08	Reclamation of areas destroyed by mining and waste dumps

Source: Demek (edt.), 1972: Manual of detailed geomorphological mapping, Academia.[15]

Different proposals for taming the river Sapt Kosi have, from time to time, been under consideration of Government. As early as in the year 1896-97 a conference was held at Calcutta (now Kolkata) presided over by the Secretary, Public Works Department, and Government of India to discuss the question of building extensive embankments to control the swinging of the river. The proposal was considered to be of doubtful efficacy and it was concluded that no steps were feasible for controlling the river with its numerous channels and their elevated beds beyond protecting by short lengths of embankments isolated tracks exposed to its flood.

The subject was again discussed in the 1937 Patna Flood conference. at this conference the problem centred mainly round "embankments public and private on the plea that embankments did more harm than good and that they merely transferred trouble from one area to another and they gave also a false sense of security". In 1953 scheme was formulated on larger scale and estimated to cost Rs. 45.90 crores.

Benefits anticipated to be received after the completion of the project are as below:

a) Flood protection for the Kosi ravaged area from Hanumannagar to Bangaon.

b) Prevention of westward movement of Kosi and consequent

destruction of buildings and cultivated lnads.

c) Irrigation facilities to 14 lac acres of land on the eastern side.

d) Irrigaiton at a later stage to 6 lach of acres on the western side of river.

e) Easy all-weather communication between eastern and western districts of Kosi.

f) Avoiding Government expenditure on flood relief measures in North Bihar.

g) Additional food production of 2.26 lacs of tons and jute and sugarcane to the extent of 3.8 lacs of tons.

h) Generation of 5,000 k.w. of power.

SAPT KOSI HIGH MULTIPURPOSE PROJECT: Indo Sapt Kosi High Multipurpose Project, Indo-Nepal Preliminary studies of Sapt Kosi High Dam Multipurpose Project envisages construction of a 269 m high dam to divert river waters through a dam toe power house with an installed capacity of 3000 MW (at 50 % load factor) and irrigation of 15.22 lakh ha Gross Command Area through construction of a barrage 1 km downstream of the dam. An additional capacity of 300 MW is further contemplated by construction of three canal type power houses along the canal system. Field investigation studies and preparation of DPR for Sapta Kosi High dam Multipurpose Project and Sun Kosi Storage-cum-Diversion Scheme are to be taken up jointly by Govt. of India and Nepal.

SUN KOSI STORAGE-CUM-DIVERSION SCHEME: Indo Diversion Scheme, Indo-Nepal Based on the preliminary studies carried out so far, four alternatives proposals for present study by Joint Project Office (SKSK) have been selected for Sun Kosi Storage-cum-Diversion Scheme. An optimal option amongst the four alternatives is required to be investigated in detail. For taking up the field works, the requirement of topographical survey and geo-technical investigations for the selected layout is to be finalized. (Annual Report-2009, central water commission, India. p. 34)

Table - 12.5
KOSI CANAL PROJECT

S. N.	Project Name	State	Major/ Medium	Estimated Cost	Benefit
				(Rs.Cr.)	in Ha./MW
01	Eastern Kosi Canal Project (New ERM)	Bihar	Major	Rs/750.75	735500
02	Extension of Embankment on left and Right bank of River Kamala Balan in lengths of 11.42 km and 5 km with Brick soiling road on top in left over reaches and protection work at two points on extended portion of right Kamla Balan embankment, Bihar	Bihar	Flood Control	56.12	3,16,000

Source: Annual Report-2009, central water commission, INDIA. P. 51 (http://wrd.bih.nic.in/west_kosi.html).

A critical evaluation of the flood control measures should be immediately undertaken and a document prepared containing the historical performance of the flood control strategies, impacts of the existing flood control measures, knowledge gaps, reports from the specific sites. It should also outline the data required for taking up detailed studies, alternative schemes for flood mitigation and identify the areas in order of priority. It is important to take up integrated basin studies - 'source to sink' and data required should be made available to scientists and engineers to propagate alternative schemes of flood control (e.g. rainwater harvesting, artificial recharge, canal system, dredging, river training, interlinking etc.) and drainage control (S., Rajiv, 2008,).

BIBLIOGRAPHY

1. Ackers, P and White, W.R., 1980: Bed material transport, A theory of for total load and its verification, Proc., 1st Intern. Symposium on river sedimentation, Beijing, China, March, pp. 110-120.

2. Ahmad, E., 1962: Geomorphic regions of Peninsular India, Jour. Ranchi Uinv., Vol.I, pp. 1-20.

3. Ahmad, E., 1965: Bihar: A physical economic and regional Geog., R.U., Ranchi, pp. 309-328 & p. 27.

4. Ahmad, E., 1968: Distribution and causes of gully erosion in India – Selected papers, Vol.1, 21st International Geog. Cong., p. 3.

5. Ahmad, E., 1979: Geomorphology and communications, Fire lectures on applied Geog., pp. 18-27.

6. Ahmad, E., 1985: Geomorphology, Kalyani Publisher, Calcutta, pp. 67-70.

7. Agarwal, C., & Sharma, V., 2012: Research Methodology in Geography, Commonwealth, pp. 73-100.

8. Agarwal, R. P. and Bhoj R., 1992: Evolution of Koshi river fan, India: structural implications and geomorphic significance, International journal of remote sensing 13 (10), pp. 1891–1901.

9. Agnihotri, S.P., Singh, P.K. and Agnihotri, A., 2009: River channel morphology and flood plain response to large-scale engineering works, Geomorphology in India, (Eds) H.S. Sharma and V.S. Kale, Prayag Pustak Bhavan, Allahabad, pp. 69-84.

10. Alford, D. and Armstrong, R., 2010: The role of glaciers in stream flow from the Nepal Himalaya, The Cryosphere

Discussions, Vol.4, pp. 469-494.

11. Anderson, R.S. & Anderson, S.P., 2011: Geomorphology: The mechanism and chemistry of landscape, Cambridge University Press, pp. 380-475.

12. Assine, M.L., 2005. River avulsions on the Taquari megafan, Pantanal wetland, Brazil. Geomorphology, pp. 70, 357–371.

13. Bajracharya, S.R. and Joshi S., 2001: Risk assessment of Tsho Rolpa Glacial Lake along the Rolwaling and Tama Koshi Valleys, Dolakha District. United Nations Environmental Programme, Regional Resource Center - APN (UNEP/RRC-AP) and ICIMOD, p. 64.

14. Bapalu, G.V. and Sinha, R., 2005: GIS in flood hazard mapping: a case study of Kosi river basin, India GIS@Development online edition, Oct., pp. 1-3.

15. Bandyopadhyay, M.K., 2008: Garhwal Himalaya – a glaciologist's perspective, 'Mount Nilkanth 6596 m' (celebrating the west ridge ascent), M.H. Mehta and P. Gupta (eds.), Souvenir, Himalayan club, pp. 5-12.

16. Barry, R.G., Chorley, 1968: Atmosphere, weather and climate, Routledge, Taylor and Francis gropus, London and Newyork, pp. 262-289.

17. Banerjee, D.K., 1997, Rivers of India, pp. 51-54.

18. Basu, S., 1976: Some Fundamental problems of meander formation with special reference to the Bhagirathi river, Geog. Rev. India, Vol. 38, No. 1, pp. 54-59.

19. Basu, S.R. & Ghatowar, L., 1988: Landslide in the Lish-Gish basins of the eastern Himalaya, Landscape System and Ecological Studies, Calcutta, Vol. 11, June, pp. 47-48.

20. Basyal, G.K. and Khanal, N.R., 2001: Process and characteristics of urbanisation in Nepal, CNAS Jourl., Vol. 28, No. 2, pp. 188-214.

21. Bhusal, J.K., 1998: Sediment transport on major rivers in Nepal, Aspects and impacts of a changing sediment

regime. Asian Institute of Technology, Bankok, Thailand, Nov 16-20-1998, pp. 31–38.

22. Bhusal, J.K., 2008: Hydrology of Solu Khola (River) : A tributary of Dudh Kosi in Kosi River Basin, Nepal, Society of hydrologists and meteorologists Nepal, (SOHAM Nepal), pp. 1-4.

23. Bilham, R., Gaur, B.K., and Molhar, P., 2001, Himalayan seismic haszard, Science pp. 1442-1444.

24. Bishop, W.W., 1958: Straitigraphical Geomorphology, Essays in Geomorphology, Edt. by G.H. Dury, Heinemann, London, pp. 139-176.

25. Blaszczynski, J.S., 1997: Photogrammetric Engineering & Remote Sensing, Ame. Soc. for Photogrammetry and Remote Sensing, Vol. 63, No. 2, pp. 183-191.

26. Bloom, A.L., 1978 (Reprint-2012): Geomorphology: A systematic analysis of late Cenozoic landforms, Rawat Publication, New Delhi, p. 3 & pp. 199-200.

27. Bloom, A.L., 1978: Geomorphology, Prentice-Hall of India, Pvt. Ltd., New Delhi, pp. 278-280.

28. Bloom, A. L., 2004 (3rd Edt.): Geomorphology: A systematic analysis of late Cenozoic landforms, Rawat publications, New Delhi, pp. 198-229, 237 & 252.

29. Bose, S.C., 1967: Geomorphology of the glaciated regions of the western Himalaya Proc. Geom. St. Ind., Univ. of Saugar, pp. 19-29.

30. Brice, J.C., 1962: Stream Laboratory Studies in Geology, No. 207, Sanfrancisco & London, W.H. Freeman & Company, p. 68.

31. Brown, R.H., 2002: Groundwater studies; studies and reports. In Hydrology 7, UNESCO, Paris, pp. 67-70.

32. Brunden, D., 1980: 'Weathering in process in Geomorphology, Embleton C & J Thornes (edt.), Arnold, Heinemann publisher, New Delhi, India, pp. 73-129.

33. Bruun, P., 1963: 'Engineering aspects of sediment transport', In review in engineering geology, The Geol. Soc. of Am., Vol. 1, pp. 39-103.

34. Brush, L.M., 1961: Drainage basins, channels and flow, characteristics of the selected streams in the central Pennsylvania, U.S. Geol. Survey, refereed paper, p. 288F.

35. Budel, J. (1963). Klima-genetische Geomorphologie, Geogr. Rundschau, 7, pp. 269-286.

36. Budel, J., 1982: Climatic geomorphology, Tran. From German by Fischer, L., and Busche, D., Princeton Univ. Press, Princeton, N.J., p. 443.

37. Campbell J. G., 2005: Inventory of Glaciers, Glacial Lakes and the Identification of Potential Glacial Lake Outburst Floods (GLOFs) Affected by Global Warming in the Mountains of India, Pakistan and China/Tibet Autonomous Region, APN & ICIMOD, pp. 20-23.

38. Carthaigh, M. M., 1989: A statistical analysis of river flow, The eastern water resource region, Water resource section, Environmental research unit, Ireland, pp. 1-3.

39. Chakraborty, T., Kar, R., Kar, P. & Basu, S., 2009: Kosi megafan: Historical records, geomorphology and the recent avulsion of the Kosi River, Elsevier, Quaternary International, 227, pp. 143-160.

40. Chandana, R.C., 2006, A Geography of Population, Kalyani publisher, New Delhi, pp. 47-48.

41. Chatterjee, S.C., 1959: The Physical basis of Geography, Ram narain lal, publ. & bookseller, pp. 115-123.

42. Chatterjee, S.P., 1940: Geomorphological mapping, Proc. Geom. St. India, Univ. of Saugar, pp. 177-187.

43. Chatterjee, S.P., 1965: Geography at International level, Geographical outlook, Ranchi, Vol. IV, p. 5.

44. Chikita, K., Yamada, T., Sakai A., Ghimire, R.P., 1997: Hydrodynamic Effects on the Basin Expansion of Tsho Rolpa Glacier Lake in the Nepal Himalaya. In Bull. of

Glacier Research (Data Center for Glacier Research, Japanese Soc. of Snow and Ice), Pub. No. 15, pp. 59-69.

45. Cheetam, G. H., 1976: Paleohydrological investigation of river terrace model. In Geomorphology (Eds. Davidson, D. A. & Shakpley, M. L.), Duckworth, London, pp. 335-344.

46. Chen, N.Sh., Hu, G.Sh., Deng, W., Khanal, N., Zhu, Y.H. and Han, D., 2013: 'On the water hazards in the trans-boundary Kosi river basin', published by Natural Hazards and Earth System, Sciences, pp. 796-797 & 805.

47. Chorley, R.J., 1961: The drainage basin as a functional geomorphic unit, In water earth and man (Edt. by R.J. Chorley), Matan, London, pp. 77-100.

48. Chorley, R.J., 1985: Geomorphology, Methuen and Co. Ltd., London, Newyork, p. 216.

49. Clark, J.I., 1966: "Morphology from Maps" An essays in Geomorphology, Edt. by G.H. Durry, Heinemann, London, p. 37 & 235-274.

50. Clozier, R., 1940: 'Les causes de Quercy', Paris, Bailliere, p. 183.

51. Crag, R.G. and J.L. Craft (Edt.), 1982: Applied geomorphology, Allen and Unwin, London.

52. CWC, 2009: General guidelines for preparing river morphological reports, Ministry of water resources, Govt. of India, New Delhi.

53. CWC, 2012: Integrated Hydrological Data Book, (Non-classified river basins), Hydrological data directorate information systems organisation, Water planning & Projects wing, New Delhi, pp. 156-163.

54. Das, K. K. L. & Das, K. N., 1981: Alluvial Morphology of the North Bihar Plain – A study in applied geomorphology, in Sharma, H. S., Perspectives in geomorphology 4, New Delhi, Naurung Rai Concept Publishing Company, pp. 85–105.

55. Das, P.K., 1968: The Monsoon, National book trust, India, pp. 181-216.

56. Das, S.K.N., 1968. Soil erosion and the problem of silting in the Kosi catchment. Journal of Soil and Water Conservation in India 16 (3–4), pp. 60–67.

57. Davis, W.M., 1909: Geographical assey, Ginn and Co Company, Boston, London, pp. 67-100.

58. Dayal, P., 1968: The Bihar plain- A regional study, Transaction, Indian council of Geog. – I.G.U., Vol. 5, p. 23.

59. Demek, J., 1982: Geomorphological Mapping: Progress and Problems in Perspectives in Geomorphology, Edt. H.S.Sharma, Concept International Series in Geography, No-2, New Delhi.

60. Devkota, L., Crosato, A. and Giri, S., 2012: Effect of the barrage and embankments on flooding and channel avulsion case study Koshi River, Nepal, A Journal of Rural Infrastructure Development, Society of Engineers' for Rural Development, Nepal (SERDeN), Vol 3. Issue 3, pp. 124-132.

61. Dhanya, V., 2014: Evolution of Achankovil drainage pattern: A geomorphic analysis, Annals, NAGI, Vol-XXXIV (No. 1), June, pp. 59-70.

62. Dhinwa, P.S., Pathak, S.K., Sastry, S.V.C., Rao, M. Majumdar, Chotani, K.L., Singh, M. L., Sinha, R.L.P. and J.P., 1992: Land use change analysis of Bharatpur district using GIS, Photonirvachak, Jour. Indian Soc. Remote Sensing, Vol. 20(4), pp. 237-250.

63. Dijkshoorn, K. and Huting, J., 2009: Soil and Terrain database for Nepal, ISRIC, World Soil Information, pp. 10-11.

64. Dingman, S.L., 1994: Physical hydrology, Macmillan Publising Company, New York, p. 575.

65. Dixit, A., 2009: Kosi Embankment Breach in Nepal: Need for a Paradigm Shift in Responding to Floods, Econ. Polit. Weekly, 7, pp. 70–78.

66. Dunne, T., 1988: Geomorphologic contributions to flood control planning, in Flood Geomorphology, edited by V. R. Baker, R. C. Kochel, and P. C. Patton, , John Wiley, New York,. pp. 421–438.

67. Dury, G.H., 1969: Hydraulic geometry, in water earth and man, Edt. by R.J. Chorley, Methun & Company, London, pp. 319-330.

68. Dury, G.M., 1982: "Methods of Cartographical Analysis in Geomorphological volume, Madras, pp. 136-140.

69. Egyed, L., 1957: 'The role of tectononics and morphology in the development of drainage pattern, Gerlands Beitr., Geophysics, 66, pp. 271-273.

70. Einstein H.A., 1978: On large scale bar pattern and channel, Proc. 22nd Japanese conference on hydraulics.

71. Fairbridge, W.R., 1968: The Encyclopedia of Geomorpholgy, Encyclopedia of earth sciences series vol. 3., Reinhold book corp., New York, pp. 282-283, 284 -285, 904 & 958.

72. Gaddes, A., 1960: "The alluvial morphology of India-Gangatic plain; Transactions and Papers, Inst. Of Br. Geographers, (contour in the map of the plain are drawn at 10' interval), 262.

73. Gao, J., Masson-Delmotte, V., Risi, C., He, Y., and yao, T., 2013: What controls precipitation $d^{18}O$ in the southern Tibetan Plateau at seasonal and intra-seasonal scales? A case study at Lhasa and Nyalam, IMI in Stockholm, pp. 5-9.

74. Garde, R.J., 1995: History of fluvial hydraulics, New Age International Pvt. Ltd., Publishers, New Delhi, pp. 4-5.

75. Garde, R.J., 2006: River Morphology, New age international (P) Ltd., publishers, pp. 15-16, 24, 120-121, & 396-403.

76. Garrels, R. M. and Mackenzie, F. T., 1971: Evolution of sedimentary rocks: W. W. Norton & Co., Inc., New York, p.

44 & 397.

77. Ghimire, S.K., Higaki, D., and Bhattarai, T.P., 2006. Gully erosion in the Siwalik Hills, Nepal: estimation of sediment production from active ephemeral gullies. Earth Surface Processes and Landform, v. 31., pp. 155–165. http://dx.doi.org/10.1002/esp.1320

78. Ghosh, M., 1998: Ph.D. Thesis, Fluvial Dynamics of the Kalikhola basin, R.U., Ranchi, pp. 174-194.

79. Ghosh, S. 1970: 'Physical and economic factor in population distribution in Bihar, The National Geog., Journ., of India, Vol. 16., pp. 61-70.

80. Gilbert, G.K., 1876: The Colorado plateau province, as a fluvial for geological studies, Amer. Jour. Sci., Vol.-12, pp. 16-24 & 85-103.

81. Gilbert, G.K., 1909: The convexity of hillslops, Joul., Geol., Vol. 17, pp. 344-351 & 137-141.

82. Gohain, K., & Parkash, B., 1990: Morphology of the Kosi Megafan. In: Alluvial Fans: a Field Approach (Ed. A.H. Rachocki and M. Church), John Wiley, Chichester, UK pp. 151-178.

83. Gragrey, K. J., 2005: Background to paleohydrology, A Whille, Interscience publication, Jhon Wille & Sons, Chicester, pp. 167-170.

84. Gupta, A., Kale, V.S. and Rajaguru, S.N., 1999: The Narmada River, India, Through space and time, In Miller, A.J. and Gupta, A., (Eds), Varieties of fluvial form, Newyork, John wiley and sons, pp. 113-143.

85. Gurjar, R.K., and Jat, B.C., 2012: Geography of water resources, Rawat publications, New Delhi, pp. 238-258.

86. Guymon, G.L., 1994: Unsaturated zone hydrology, PTR Prentice Hall, Englewood cliffs, New Jersey, pp. 32-55.

87. Hack, J.T., 1975: Dynamic equilibrium and landform evolution, In theories of landform development, Edt. By W.N. Methorn, R.C. Flaemal, George Allen and Unwin, pp.

67-70.

88. Hagging, C.G., 1975: Dynamic equilibrium and landscape evolution, In theories of landforms, Edt. By W. N. Melhorl & Flamal, Gorge Allen & Unwin, p. 67.

89. Hanumantha & Rao, C.H., 2000: "Watershed Development in India: Recent Experience and Emerging Issues", Economic and political weekly, Vol-35(45), pp. 3943-3947.

90. Hasnain, S. I., 2002: Himalayan glaciers meltdown: impact on South Asian Rivers. IAHS Pub No. 274, pp. 1-7.

91. Hausler, H.; Leber, D.; Buchroithner, M. F., 1993: Geoecological studies using multitemporal multisensor remote sensing data in the "Three River Valley Area" Southern Xizang (Tibet), China. Proc. 2nd Int. Symp. on High Mountain Remote Sensing Cartogr., Astronautical Publ. House, Beijing, pp. 106-119.

92. Hein & Gansser, 1939, Geological structure of the central Himalaya, Swiss natural science society, mem., vol. XXIII, p. 380.

93. Hetzel, R., Dunkl, I., Haider, V., Strobl, M., Von, E. H., Ding, L. and Frei, D., 2011: Peneplain formation in southern Tibet predates the India-Asia collision and plateau uplift: *Geology, vol.* 39, *pp.* 983–986.

94. Hicking, E.J., and Nanson, G.C., 1975: The character of channel migration on the Beaton river, north east British Columbia, Canada, Bull. Geol. Soc. Amer. Vol-86, pp. 487-494.

95. Higgins, C.G., 1980: Applied Climatology, Butterworths, London, pp. 487-494.

96. Highway Research Board, 1958: Speed Report-29, Landslide and engineering practice, Washington, p. 232.

97. Hire, P.S. and Kale, V.S., 2009: Dynamics of suspended sediment transport in the Tapi basin, India, Geomorphology in India, (Edt) H.S. Sharma and V.S. Kale, Prayag Pustak havan, Allahabad, pp. 101-120.

98. Holmes, A., 1963: Principles of Physical Geology, Second edition, Thomas Nelson and son Ltd., London, pp. 473.

99. Holmes, R.R., 2010: Measurement of Bedload Transport in Sand-Bed Rivers: A Look at Two Indirect Sampling Methods, U.S. Geological Survey Scientific Investigations Report 2010-5091, U.S. Geological Survey, Rolla, Missouri, United States, pp. 236-252.

100. Hortons, R.E., 1933: Drainage basin characteristics, Trans. Amer. Geophysics. U., Vol-14, pp. 350-361.

101. Hortons, R.E., 1945: Erosional development of stream, their drainage basin: Hydrological approach to quantitative morphology, Bull. Geol. Soc. Am., Vol. 50, pp. 130-140, 149-150 & 275-370.

102. Horton, R.E., 1945: Erosional development of streams and their drainage basins: Hydrological approach to quantitative geomorphology, Bull. Geol. Soc. Am., Vol-56, pp. 310 & 350.

103. Hussain, M., 2009 (Reprint-2010): Fundamentals of physical Geography, Rawat publications, New Delhi, p. 220.

104. IPCC, 2012: Managing the risks of extreme events and disasters to advance climate change adaptation, A special report of the IPCC, Cambridge University Press, p. 29.

105. IPCC, Climate change 2013: The physical science basis is the contribution of Working Group I, Cambridge University Press, p. 1217.

106. Jain, M.K., Kothyari, U.C., and Ranga Raju, K.G., (2004), A GIS based distributed rainfall – runoff model, Journal of Hydrology, pp. 299 (2004), 107-135.

107. Jain, S. K.; Agarwal, P. K. and Singh, V. P., 2007: Hydrology and water resources of India, Dordrecht, Springer. pp. 358–359.

108. Jain, V. and Sinha, R. (2003a): Derivation of Unit hydrograph from GIUH analysis for a Himalayan river.

Water Resources Management, Vol.17, pp. 355-375.

109. Jain, V. and Sinha, R. (2003b) Evaluation of geomorphic control on flood hazard through GIUH. Current Science, v. 85 (11), pp.1596-1600.

110. Jain, V. and Sinha, R., 2003c): Geomorphic Manifestation of the flood hazard: A remote sensing based approach, Geocarto International, 18, 4, pp. 51-60.

111. Jha, R., 2008: Total Run-of-River type Hydropower Potential of Nepal, pp. 1-15.

112. Jha, S., 2009: Regional Geography of Kosi-Mahananda Interfluve, Janaki prakashan, Patna, pp. 93-100.

113. JICA 1987: Draft Final Report on Feasibility Study on Arun 3 – Hydro-Electric Power Development Project Vol. I Main Report, pp. 5-43.

114. Johnson, D.W., Dover Publication, INC., (Edt.) 1990: The peneplain, originally written in reply to a paper by Prof. R. S. Tar on the same subject. Reprinted with numerous minor changes. Am. Geol., Vol. 23, pp. 207-239, also in Geographical Essays, W. M. Davis, pp. 350-380.

115. Jones, D.K.C., 1980: British applied geomorphology: An appraisal, Zeitschrift fur Geomorphologie, supplement 36, pp. 48-73.

116. Jordan, G., 2003: Morphometric analysis and tectonic interpretation of digital terrain data: A case study of surface processes and landforms, Vol. 28, pp. 807-822.

117. Jones, D.K.C., 1980: British applied geomorphology: An appraisal, Zeitschrift fur Geomorphologie, supplement 36, pp. 48-73.

118. Jones, L.S. and Schumm, S.A. (1999) Causes of avulsion: An overview. In: N. D. Smith and J. Rogers (eds.) Fluvial Sedimentology VI, Special Publication of International Association of Sedimentologists, Vol. 28, pp. 171-178.

119. Kaldova, Jan, 1972: Geomorpholocal study in the Himalaya, special reference to landslide and allied

phenomena in Himalayan Geol. Vol. 2, Eds. A.G. Jhingran, K.S. Wadia & A.K. Jain, Wadia Inst. Of Himalayan Geol., Delhi, pp. 78-80.

120. Kale,.V. S, & Gupta, A., 2001: Introduction to Geomorphology, Universities press.

121. Kale,.V. S, 2002: Fluvial geomorphology of Indian rivers; an overview. Progress in Physical Geogphay 26, pp. 400-433.

122. Kale,.V. S. & Hire, P. S., 2004: Effectiveness of large monsoon floods on the Tapi river, India: role of channel geometry and hydrologic regime. Geomorphology p. 57 & pp. 275-291.

123. Kale, V. S., Singhvi, A. K., Mishra, P. K. and Banergee, D., 2000: Sedimentary records and Luminescence chorology of late Holocene palaeo floods in the Luni river, Catena, Elsvier publishers, Vol. 40, pp. 337-358.

124. Kapoor, B.L., 1967: Application of Geomorphology in forecasting military potential in inaccessible areas, Geom. Stud. In Ind. Proc., Sagar, p. 201.

125. Karki, A., 2005: Indrawati River Basin Study Draft Report, Consolidated Management Services Nepal (CMS) Min Bhawan, Bansewor, Kathmandu, pp. 03-04.

126. Kattelmann, R., 1990: "Hydrology and development of the Arun River, Nepal", Proceedings of two Lausanne Symposia, August 1990, IAHS publ. no. 193. Inter. Assoc. of Hydro. Sci. p. 778.

127. Kattelmann, R., 1991: Hydrologic Regime of the Sapt Kosi bsin, Nepal, pp. 139-145.

128. Kaushic, S.D., 1967: Application of Geom. In terrain-evolution & lnad utilization in India, Proc. Geom. Stud. India, p. 200.

129. King, C.A.M., 1966: Techniques in Geomorphology, Edward Arnold, London, pp. 166-170.

130. King, L. C., 1957: The uniformitarian nature of hillslope, Trans. Royal Geol. Society, Edinburgh, Vol. 17, pp. 81-102.

131. Kothyari, U. C. & Garde, R. J., 1991: Annual runoff estimation for catchment in India, JWRPM, Proc. ASCE, Vol. 117, No. 1, pp. 1-10.

132. Krishanan, M.S., 1982 (Reprinted 2011): Geology of India and Burma, BBS pulisher & distributer pvt. Ltd., New Delhi, pp 138-139.

133. Krumbein, W.C. & Sloss, L.L., 1953: Stratigraphy and sedimentation, W.H. Freeman and company, San Francisco, pp. 66-83, 148-186 & 186-223.

134. Kumaaswamy, K., Moses, A.A., & Vasanthy, M., 2004: Environmental studies, Bharathidasan University, Triuchirappalli; pp. 219-224.

135. Lal, A.K., 1998: Morphometric determinents of drainage cycles in Nayar basin, Garhwal Himalaya, Hima. Geol., 19 (1), pp. 55-68.

136. Lal, D.S., Climatology, 2012: Sharda pustak bhawan, pp. 392-402.

137. Lai, V. B., Banerji, S. and Narayanan, J., 1970: Sediment yield in relation to drainage basin characteristics in some Indian river valley projects, pp. 93-97.

138. Leber, D. M., Holawe, F. & Hdusler, H., 1995: Climatic Classification of the Tibet Autonomous Region Using Multivariate Statistical Methods, GeoJournal, Kluwer Academic Publishers, 37.4, pp. 451-473.

139. Leopold, B. Luna, Wolman, M. Gordon and Miller, P. John, 1995: Fluvial processes in Geomorphology, pp. 198-266 & 333.

140. Leopold, L.B. & Maddock, T., 1953: The hydraulic geometry of stream channels and some physiographic implications, U.S. Geol. Surv. Prof. Papers. p. 252.

141. Leopold, L.B. and Woldman, M.G., 1957: River channel patterns, U.S. Geol. Survey, Paper- 282-B, reprinted in

rivers and river terraces, ed. by G.S. Duvry, Mechmilon, 1970, p. 211.

142. Leopold, L.B., Woldman M.G. & Miller, J.P., 1969: Fluvial processes in Geomorphology, W.H. Freeman, San Francisco, pp. 150-160.

143. Leopold, L.B., 1953: Downstream change velocity in rivers; Am. Jourl. Sci., Vol. 251, pp. 606-624.

144. Leopold, L.B. & Wolman, 1960: River Meanders, Bull. Geol. Soc. Am., Vol. 71, p. 350.

145. Leopold, L.B., Woldman, M.G. & Miller, J.B., 1964: Fluvial processes in Geomorphology, Eurasia publishing house, New Delhi, p. 143.

146. LIGG/WECS/NEA, 1988: Report on First Expedition to Glaciers and Glacier Lakes in the Pumqu (Arun) and Poique (Bhote-Sun Kosi) river basins, Xizang (Tibet), China, Sino-Nepalese Investigation of Glacier Lake Outburst Floods in the Himalaya. Beijing, China: Science Press.

147. Liu, C. H. and Sharma, C. K., 1988: Report on First Expedition to Glaciers and Glacier Lakes in the Pumqu (Arun) and Poiqu (Bhote-Sun Kosi) River basins, Xizang (Tibet), China, Science Press, Beijing, pp. 1–192.

148. Liu, W., 2006, Prebinary study on debris flow induced by glacier lake outburst in Tibet, Hydrology and Engineering geology, vol. 3, pp. 88-92.

149. Li Zhen, Sun Wenxin, Zeng Qunzhu, 1998: Measurements of glacier variation in the Tibetan Plateau usingLandsat data. Remote Sensing of Environment, 63(3), pp. 258-264.

150. Lubowe, J.K., 1964: Stream junction angle in the drainage pattern, Am. Journ. of Sci., Vol-262, pp. 325-339.

151. Lutgens, F.K. & Tarbuck, E.J.: 2012: The atmosphere – An introduction to Meteorology, PHI learning Pvt. Ltd., New Delhi, pp. 416-426.

152. Mahato, R.K., 2008: Morphometric analysis of “Sapahi

River", unpublished M.A. thesis, Ranchi University, pp. 25-55.

153. Mandy, J., Stasiuk, M., Shaw, J., Darren, B. S., Tracy, A. B., Timothy, G. F., David, R. S., Phillip, S. G. K., Claire, L., Rains, B. R. and Bruce, B. R., 2009: The morphology and sedimentology of landforms created by subglacial megafloods, Megaflooding on Earth and Mars, ed. Devon M. Burr, Paul A. Carling and Victor R. Baker, Published by Cambridge University Press, pp. 78-103.

154. Melton, M.A., 1958B: Correlation structures of morphometric property of drainage system and their controlling agents, Jourl. of Geol., Vol-66, pp. 35-56.

155. Melton, M.A., 1966: Geomorphic properties of Mahire drainage system and their representation in an E4 space, Joul. of Geol., pp. 35-54.

156. Mishra, J.P., 1967: Structural control on the evolution of the Himalayan river system, Proc. Geom. St. India, pp. 87-91.

157. Mittal, R.S., Prakash, B. and Bajpai, I.P., 1974: Drainage basin morphometric study of the part of Garhwal Himalaya, Himalayan Geology, Wadia Institute of Himalayan Geology, pp. 67-70.

158. Moglen, G.E., Eltahir, E.A.B. & Bras, R.L., 1998: On the sensitivity of drainage density in climate change, Water resource research, Vol-34, No. 4, pp. 855-862.

159. Mool, P.K., 1998: Use of Multi-Temporal Data for the Study of Glacier Lakes and Glacier Lake Outburst Floods in Nepal Himalaya: Tsho Rolpa Glacier Lake as a Case Study. In Proceedings of International Symposium on Application of Remote sensing and Geographic Information System to Disaster Reduction, Tsukuba, Japan, 3 March 1998, pp 13–21.

160. Montgomery, D.R., 2001: Slope distributions, threshold hillslopes and steady-state topography. American Journal of Science, pp. 301, 432–454.

161. Montgomery, D.R., & Stolar D.B., 2006: Reconsidering Himalayan river anticlines, Elsevier, Geomorphology, Vol.-82, pp. 4-15.

162. Moore, W.G., 1993: A dictionary of Geomorphology, Penguin books Ltd., London, Middlex, p. 166.

163. Morisawa, M., 1975: Tectonics and Geomorphic models, In theories of landform development, Edt. By W.N. Melhorl, R.C. Flemal, George Allen and Unwin, London, pp. 199-216.

164. Morisawa, M.E., 1977: Fluvial Geomorphology, S.U.N.Y.-Brighton, Newyork, pp. 9-21.

165. Morisawa, M., 1985: Rivers (Forms and Processes), Longman, London and Newyork, pp. 40-47, 95-100 & 185-190.

166. Mukhopadya, S.C., 1982: The Tista basin-A study in fluvial geomorphology, K.P. Bagchi and Co., Kolkata, pp. 308.

167. Nag, P. and Saha, G.N., 1996: Edt. Geomorphological Mapping, Perspectives & Dimensions, NATMO, pp. 5-8 & 13-22.

168. Nash, D.B., 1994: Effective sediment-transporting discharge from magnitude-frequency analysis, The jourl. of the Geol., Vol-102, pp. 79-95.

169. Naithani, A.K., Nainwal, H.V., Sati, K.K. and parsad, C., 2001: Geomorphological evidences of retreat of the Gangotri glacier and its characteristics. Current Science 80(1), 10 January 2001. pp. 87-94.

170. Narayan, H., 1965, "Year borne magnetic survey, Proceeding of seminar of Earth Sciences, PT. 1-Geophysics, the Indian Geophysical Union, pp. 190-195.

171. Nayak, J.N., 1996: Sediment management of the Kosi River basin in Nepal, Erosion and Sediment Yield: Global and Regional Perspectives (Proceedings of the Exeter Symposium July 1996). IAHS Publ. no. 236, pp. 583-586.

172. NRSC and CWC, 2011: Assessment of water resources at

basin scale using space input, A pilot study of Godavari and Brahmani basins, Water resource division, NRSC, ISRO, Hyderabad, pp. 35-63.

173. Owen, L.A., 1995, Shaping the Himalayas, Geological Magazine, vol. 12, pp. 23-25.

174. Pal, S.K., 1972: A classification of morphometric methods of analysis: An appraisal, Geog. Rev. India, Vol. 34, No. 1, pp. 61-84.

175. Panda, P.K., 2012: Hydrological and hydrodynamic modeling of Kosi river paleochannel, Bihar (India), International Journal of Darshan Institute on engineering research and emerging technology, Vol. 1, No. 1, pp. 52-58.

176. Pannell, C.W. and Ma, L.J.C., 1983: China, The geography of development and modernization, V.H. Wilson & Sons, Edward Arnold, pp. 7-27.

177. Parksash, B. & Kumar, S., 1991: The Indo-Gangetic Basin: Sedimentary Basins of India, tectonic Context (Edt. By S. K. Tandon, C.C. Pant and S.M. Casshypa), Gyanodaya Prakashan, Nainital, India, pp. 147-170.

178. Pascoe, E., 1964: A manual of geology of India and Burma, Vol. 3, Manager of Publication, Joul. of India, p. 67.

179. Peltier, L. C., 1950: The geographic cycle in periglacial regions as it is related to climatic geomorphology, Ann. Ass. Am. Geographers, Vol. 40, pp. 214-236.

180. Petes, G.N. & Foster, I., 1985: River and Landscape, Edward Arnold, pp. 67-70 & 578.

181. Prakash, A, 1978: Geomorphology of Lawa river basin, Ph.D. thesis (unpublished), pp. 440-443.

182. Parkash, B. & Kumar, S., 1991: The Indo-Gangetic Basin: Sedimentary Basins of India, Tectonic Context (Ed. By S.K. Tandon, C.C. Pant and S.M. Casshypa), Gyanodaya Prakashan, Nainital, India, pp. 147-170.

183. Prasad, G. & Nautiyal, R., 2012: Climate and global

warming, Sharda pustak bhawan, Allahabad.

184. Prasada, R., P.V.S.P., 1998: Morphology and evolution of an alluvial fan – A case study of Katanutala, Prakashan district, Joul. of the Indian Soct. Of Remote Sensing, Vol. 16, pp. 21-30.

185. Rehman, Z. A., RAO, L. A. K. and Saran, S., 2013: Effect of Geology and Geomorphology on Landuse/Landcover in Himalayan Foothill, Dehradun, Jour. Geol. Soc. of India, Vol.81, June, pp. 827-834.

186. Reynolds, A.J., 1965: Waves on the erodible bed of an open channel, JFM, Vol. 22 pt. 1, pp. 113-133.

187. Ritter, D.F., 1978: Process Geomorphology, Wm. Co Brown publ., Dubuque, Iowa, pp. 154-302.

188. Rizvi, S.M.H., 2008: Geomorpholgy and hydrogeology: A handbook, CBS publishers & distributors Pvt. Ltd., New Delhi, pp. 184-273 & 201-216.

189. Roy, A.K., 1966 (reprint-2009): Introduction to Geological maps and structures, world press, pp. 70-112.

190. Rosgen, D.L., 1994. A classification of natural rivers, Cantenna, v.22, pp. 169–199.

191. Roy Chaudhury, P.C., 1964: Bihar District Gazetteers, Darbhanga, Printed by the Superintendent Secretariat Press, Bihar, Patna, pp. 03-04.

192. Rowland, J. C., Lepper, K., William, D. E., Wilson, C. J. and Sheldon, R., 2005: Tie channel sedimentation rates, oxbow formation age and channel migration rate from optically stimulated luminescence (OSL) analysis of floodplain deposits, Earth Surface Processes and Landforms, OSL analysis in oxbow lakes, John Wiley & Sons, Ltd. Earth Surf. Process. Landforms 30, Published online in Wiley InterScience, pp. 1161–1179 (www.interscience.wiley.com)

193. Rubey, W.W., 2008: The force required to move particals or a stream be; U.S., Geol, Survey, Prof. paper. 189-E, pp.

120-141.

194. Sarkar, G. and Gupta, S., 2009: Fluvio-Geomorphological characteristics of downstream of Panchnoi, Duars, North Bengal, India, ILEE, Vol. 32(1), June, 2009, pp. 9-14.

195. Sarkar, S. and Patel, P.P., 2009: Drainage analysis of the Dulung basin, Geomorphology in India, (Edt) H.S. Sharma and V.S. Kale, Prayag Pustak Bhavan, pp. 133-154.

196. Satpathi, D.P.P., 1981: An outline of Indian Geomorphology: A study in regional geomorphology of Sinhbhum, Classical publishing company, New Delhi, pp. 191-220.

197. Schelling, D., 1992: The tectonostratigraphy and structure of the eastern Nepal Himalaya. Tectonics 11, pp. 925–943.

198. Schumm, S.A., 1956: The effect of sediment type on the slope and the stratification of some of some modern fluvial deposit, Am. Journ. Soc., Vol. 258, p. 603.

199. Schumm, S.A., 1963. The disparity between present rates of denudation and orogeny. U.S. Geological Survey Professional Paper, vol. 454-H. United States Government Printing Office, Washington, D.C.

200. Schumm, S.A., 1975: Geomorphic threshold and complex response of drainage system, in fluvial geomorphology, Binghanton, Newyork, pp. 299-310.

201. Schumm, S. A., 1977: The fluvial system, Jhon wille and Sons, A. Willey interscience publication, Newyork, Chapter – 5, pp. 106 – 111.

202. Schumm, S.A., 1956: Evolution of drainage system and slopes in bedlands at Perth Embay, New Jersey, Bull. Geol. Soc. Am., Vol-67, p. 112, 607 & 612.

203. Schumm, S.A., 1981: Evolution of response of fluvial system, Sedimentological implications, Soct. of Economic Paleontologists and Mineralogists, Special publ., pp. 19-29.

204. Schumm, S. A. & Rea, D. K., 1995: Sediemnt yield from

disturbed earth systems, Geology, Vol. 23, pp. 391-394.

205. Seṅ, P.K. & Prasad, N., 2002: An introduction to the Geomorphology of India, Allied publishers Pvt. Ltd., pp. 110-159.

206. Seko, K., Yabuki, H., Nakawo, M., Sakai, A., Kadota, T., and Yamada, Y., 1998: Changing surface features of Khumbu Glacier, Nepal Himalayas revealed by SPOT images. Bulletin of Glacier Research 16, pp. 33-41.

207. Sharma, C.K., 1983: River systems of Nepal, published by Mrs. Sangeeta Sharma, Kathmandu.

208. Sharma, C.K., 1979: Partial drought conditions in Nepal, Hydrological Sciences Bulletin des Sciences Hydrologiques, Vol. 24, 328-333.

209. Sharma, H.S. and Kale, V.S., 2011 (Edited): Geomorphology of India, A felicitation volume (Edited), Prayag Pustak Bhavan, Allahabad, 11-17.

210. Sharma, J.P., 1983: Practical Geography, Rastogi publication, Merath.

211. Sharma, K.C., 1997: Rivers of Nepal, Nauna printing works Pvt. Ltd. 47, Ganesh Chandra Avn. Cal., pp. 27-28 (Table) & 57.

212. Shields, A.F.,1965: Hydraulic structure, Vol. 1, Am. Soc. Mech., Engineer., p. 504.

213. Shrestha, M.B., Tamrakra, N. K., and Miyazaki, T., 2008. Morphometry and sediment dynamics of the Churia River area, Siwalik Range in Nepal. Boletin de Geologia, v. 30, pp. 35–48.

214. Shrestha, P., & Tamrakar, N. K., 2012, Morphology and classification of the main stem Bagmati River, Central Nepal, Bulletin of the Dept. of Geol., T. U., Kathmandu, Nepal, Vol. 15, pp. 23-34.

215. Sinha, C.P., 2008: "Management of floods in Bihar", Economic and political weekly, Nov. 15, pp. 40-43.

216. Sinha, R., 2008: Kosi Floods: Time of retrospection, Geography and You September – October, pp. 38-39.

217. Sinha, R., 2008: Kosi: rising waters, dynamic channels and human disasters, Economic and Political Weekly, pp. 42–46.

218. Sinha, R., 1995: Sedimentology of Quaternary Alluvial Deposits of the Gandak-Kosi interfan, North Bihar Plains, Joul. Geol. Soc. of India, Vol. 46, pp. 521-532.

219. Sinha, R., 2009: Dynamics of a River System– the Case of the Kosi River in North Bihar, e-Journal Earth Science India Vol.2 (I), pp. 1-8.

220. Sinha, R. and Friend, P.F., 1994: River systems and their sediment flux, Indo-Gangetic plains, Northern Bihar, India, Sedimentology, 41, pp. 825-845.

221. Singh, S. & Ojha, S.S., 1996: Development of geomorphology in India, in New frontiers in Indian Geography, eds by L.R. Singh, R.N. Dubey, Foundation and Geography Department, Allahabad University, pp. 158-200.

222. Sinha, R., Tandon, S.K., Gibling, M.R., Bhattacharjee, P.S., and Dasgupta, 2005, A.S. Late Quaternery geology and alluvial stratigraphy of Ganga plain, Himal. Geol. Vol. 26, pp. 223-240.

223. Sinha, R., Bapalu, G.V., Singh, L.K., and Rath, B., 2007: Flood risk analysis in the Kosi river basin, north Bihar using multi-parametric approach of Analytical Hierarchy Process (AHP). Ind. J. Remote Sensing, Vol. 36, pp. 335–349.

224. Singh, H., Parkash, B., & Gohain, K., 1993: Facies analysis of the Kosi megafan deposits, Sedimentary Geol., Vol.-85, Elsevier Science Publishers B.V., Amsterdam, pp. 87-113.

225. Singh, K.N., 1968: The territorial basis of town and village settlement in eastern U.P., India,' Annals, Assoc. of Amer. Geog., 58, pp. 203-204.

226. Singh, N.K.P., 1968: "Geomorphic evolution of the northern Rupununi basin, Guyana, Dept. of Geog., Mc – Bill University, Montreal, p. 45.

227. Singh, R., Pandey, B.W. & Prasad, A.S., 2014: Extent and magnitude of flood due to soil erosion in lower Brahmaputra river basin, Assam, Annals, NAGI, Vol-XXXIV (No. 2), Dec., pp. 162-179.

228. Sinha, R., Bapalu, G.V., Singh, L.K. and Rath, B., 2008: Flood Risk Analysis in the Kosi River Basin, North Bihar Using Multi-Parametric Approach of Analytical Hierarchy Process (AHP), J. Indian Soc. Remote Sens., Vol. 36, pp. 293-307.

229. Sinha, R., & Friend, P.F., 1994, River systems and their sediment flux, Indo-Gangetic plains, Northern Bihar, India, Sedimentology, 241, pp. 825-845.

230. Singh, H., Parkash, B. and Gohain, K., 1993: Facies analysis of the Kosi megafan deposits, Sedimentary Geology, pp. 85, 87–113.

231. Singh, R., 1995, Sedimentology of Quaternary Alluvial Deposits of the Gandak-Kosi interfan, North Bihar Plains, J.G.S.I., Vol. 46, Nov, pp. 521-532

232. Singh, R.L., 1971: (Editor) India-A Regional Geography. Silver Jubliee Publication. National Geographical society of India, Varanasi.

233. Singh, R.L., 1968, Regional Geography of India (Reprint edition 2008), National Geographical Society of India, Varanasi, pp. 190-191.

234. Singh, R.P., 1977: Geomorphology and its recent trends – National Geog., Allahabad, pp. 35-45.

235. Singh, S., 1985: An approach to the study of landforms, in concepts and methods in Geography, Geography department, Burdwan University, pp. 38-50.

236. Singh, S. & Ojha, S.S., 1996: Development of geomorphology in India, in New frontiers in Indian

Geography, eds by L.R. Singh, R.N. Dubey, Foundation and Geography Department, Allahabad University, pp. 158-200.

237. Singh, S., 1998: Geomorphology, Prayag Pustak Bhawan, pp. 65, 247-248, 353-355 & 410-412.

238. Singh, S., 2003: Physical Geography, Prayag Pustak Bhawan, p. 201.

239. Singh, S., 2006 (Reprint): Geomorphology, Prayag Pustak Bhavan, pp. 336-337.

240. Singh, S., 2014: Disaster management, Prayag Pustak Bhawan, Allahabad, pp. 156-157.

241. Singh, S.C., 1965: Delemetation of Middle Ganga Plain, N.G.G.I., XIth (two zone, 1965), pp. 74-83 also changing in the courses of rivers and their effects in urban settlement in the Middle Ganga plain, "Ph.D. thesis, Dept. of Geography, B.H.U., Varanasi-1065".

242. Singh, S. & Srivastava, R., 1974: A morphometric study of the tributary basin of the upper reaches of Belan river, National Geog. Observer, Vol-12, pp. 31-44.

243. Singh, S. & Srivastava, R., 1976: Morphometric determinants of the stage of the cycle of erosion, the Geog., Observer, Vol-12, pp. 29-40.

244. Singh, S. & Upadhaya, D.P., 1982: Topological and geometric study of drainage network, South-East Chhotanagpur region, In perspective in Gemorphology, Vol-2, Concept pub. Comp., New Delhi, pp. 191-233.

245. Smith, L.C., 1997: Satellite remote sensing of river inundation area, stage, and discharge: a review, Hydrol Process, Vol. 11, pp. 1427-1439.

246. Sparks, B.W., 1960: Geomorphology, Longman Green Londar, pp. 150-152.

247. Standards for National Hydrography Dataset, 1998: High Resolution - National Mapping Program Technical Instructions. USEPA/USGS, DRAFT.

248. Stoddart, D.R., 1969: Climatic geomorphology: Review and reassessment, progress in Phy. Geog., Vol. 1. pp. 159-222.

249. Strahler, A.N., 1958: Dimensional analysis apply to fluvially eroded landform, Geol. Soc. Am. Bull., Vol-69, pp. 279-300.

250. Strahler, A.N., 1964: Quantitative geomorphology of drainage basins and channel networks, in Handbook of Applied Hydrology, edited by Chow, V.T., pp. 439-476.

251. Strahler, A.N., 1969: Physical Geography, 3rd edt., John Willi and Sons, Inc, Newyork, pp. 69-70 & 254-257.

252. Tamura, T., 1980: A note on landform classification methodology, Geogr. Paps, to the Memory of Prof. Dr. K. Nishimura & Kokon shoin, Tokyo, pp. 80-86.

253. Thornes, J.B. & Brunsden, D., 1978: Geomorphology and time, Methuen & Co Ltd., pp. 93-112.

254. Thornbury, W.B., 1957: Principles of Geomorphology, Jhon Wille and Sons, Newyork, p. 126.

255. Thornbury, W.D., 1954: (reprint 2013), Principals of Geomorphology, published by New age international (P) Ltd., pp. 20-21, 23, 36 & 37.

256. Tinkler, K.J., 1985: A short history of Geomorphology, Croom Helm (P) Ltd., U.K., 1st Edn.

257. Verstappen, H.Th., 1954: 'The influence of climatic changes on the function of coral islands', M.J.Sci., pp. 428-435.

258. Wadia, D.N., 1995: (10th reprint), Geology of India, Tata Mcgraw Hill Publishing Ltd., New Delhi, pp. 74-88.

259. Wadia, D.N., 1960: The post glacial desiccation of Central Asia, monograph, Nat. Inst. Sci. India, pp. 1-25.

260. Wager, L.R., 1937. The Arun river drainage pattern and the rise of the Himalaya. Geographical Journal, pp. 89, 239–250.

261. Wagnerk, D.L., 2002: Geomorphic Map of California prepared by Olaf P. Jennings, 1938, Revisions of the text, p. 4.

263. Walford, N., 2011: Practical statistics for Geographers and Earth scientists, Willey-Blackwelt, pp. 267-298.

262. Wells, N.A., & Dorr, J.A., 1987: Shifting of the Kosi river, northern India. Geology, 15, pp. 206-207.

263. Wells, N.A. and Dorr, J.A., 1987: A reconnaissance of sedimentation of the Kosi alluvial fan of India, In: F.C Ethridge, R. M. Flores and M .D. Harvey, (Eds.) Recent Developments in Fluvial sedimentology, SEPM, Spl. Pub. No. 39, pp. 51-61.

264. Whittow, J.B., 1984: 'The Parguin dictionary of Physical Geography', p. 88.

265. Wooldridge, S.W., and Morgan, R.S., 1960: An outline of Geomorphology, Longmans, pp. 167-170.

266. Worcester, P.G., 1948: A textbook of Geomorphology, D. Van Nostrand Co. Inc. Newyork, U.S.A. (2nd Edition), pp. 5-10.

267. WWF Nepal Program, 2005: An overview of glaciers, glacier Retreat, and subsequent impacts in Nepal, India and China, pp. 5-28.

268. Yamada, T. and Sharma, C. K., 1993: glacier lakes and outburst floods in the Nepal Himalaya, snow and glacier hydrology, pp. 28, 319-330.

269. Zernitz, E.R., 1932: Drainage patterns and their significance, Jour. Geol. 40, pp. 498-521.

270. en.m.wikipedia.org/wiki/Hydrology, 2015

271. http://www.conservation.ca.gov/cgs/information/publications/cgs_notes/note_36/Documents/note_36.pdf

272. http://en.wikipedia.org/wiki/Himalayas (accessed on 01.01.2015)

273. http://wrd.bih.nic.in/west_kosi.html

APPENDIX- I

PHOTO PLATES

APPENDIX - II
LOCAL NOMENCLATURE OF SOME RELEVANT FEATURES

S.N.	NOMENCLATURE	DESCRIPTION	REGION
1	Bagar	Eroded Upland	Pediment of Himalaya (Terai)
2	Bhabar	Dry boulder-strewn tract along he foot-hill zone of the Siwalik mountains. The rivers travelling this belt get diverted to underground passage ways leading thereby intermittent drainage pattern.	Pediment of Himalaya (Terai)
3	Bhangar (Bangar)	Older alluviam in the plains.	Pediment of Himalaya (Terai)
4	Chu	Tibetan word for water	Tibet Plateau (China)
5	Chaur	Low lying plain with semi humid land	Bihar (India)
5	Diara	Island-like lands enclosed between the bifurcated channels of a river in the plains.	Plain region of Bihar (India)
6	Doab	Land or region between two rivers.	Plain region of Bihar (India)
7	Nala	Rivulets or ravines seasonally filled in with water	Nepal & India
8	Tatopani	Hot stream	Nepal-China boarder
9	Katni	Meander side erosion	Kosi region of Bihar (India)
10	Jhils	Lakes in the riverine tracts mostly ox-bow lakes.	Kosi region of Bihar (India)
11	Kachhar	Low-laying ill-drained tracts in the plain, in some places (Kursela) people use as bottom portion's sand deposition.	Kosi region of Bihar (India)

12	Ghurna	Whirling water of river both vertically and horizontally.	Kosi region of Bihar (India)
13	Saus	Gangatic Dolphin	Bihar (India)
14	Pairo	Landslide	Nepal
15	Bhir	Scarp	Nepal
16	Suntala	Orange	Nepal
17	Kath	Wood	Nepal & India
18	Goreto/Bato	Pathways (Way for walking)	Nepal
19	Simol Tarul	Root Potato	Nepal
20	Dhulo	Dust (Fine materials in the air)	Nepal
21	Balwa	Sand	Nepal & India

Source : Collected by Researcher through extensive field survey.

APPENDIX - III
MEAN MONTHOLY SEDIMENTATION CONCENTRATION (in ppm)

S. N.	River	Location	Station No.	Year	Month	Mean monthly	Mini	Max
1	Sapt Kosi	Chatra-Kothu 26°52'00"N & 87°09'30"E	695	**1998**	June	1444	218	6505
					July	1472	596	3200
					August	1347	544	3481
					September	3053	564	33587
					October	474	100	1305
					TOTAL	7790	2022	48078
					MEAN	**1558**	**404.4**	**9615.6**
2	Sapt Kosi	Chatra-Kothu 26°52'00"N & 87°09'30"E	695	**1999**	June	1239	216	3972
					July	1751	463	4145
					August	2277	873	4774
					September	1489	430	3543
					October	1258	73	7000
					TOTAL	8014	2055	23434
					MEAN	**1602.8**	**411**	**4686.8**

3	Sapt Kosi	Chatra-Kothu 26^{0}52'00"N & 87^{0}09'30"E	695	**1998**	July	2055	1020	5460
					August	2222	443	9590
					September	6171	542	19800
					October	3703	1400	10200
					TOTAL	14151	3405	45050
					MEAN	**3537.75**	**851.25**	**11262.5**
4	Sapt Kosi	Chatra-Kothu 26^{0}52'00"N & 87^{0}09'30"E	695	**1998**	June	1704	131	6772
					July	1947	590	12349
					August	1691	886	2736
					September	998	253	2755
					October	244	49	945
					TOTAL	6584	1909	25557
					MEAN	**1316.8**	**381.8**	**5111.4**
5	Sapt Kosi	Chatra-Kothu 26^{0}52'00"N & 87^{0}09'30"E	695	**1998**	June	1299	119	3862
					July	1857	872	4970
					August	1898	438	7273
					September	1055	482	2583
					October	201	69	542
					TOTAL	6310	1980	19230
					MEAN	**1262**	**396**	**3846**

6	Dudhi Kosi	Rabuwa Bazar 27°16'00"N & 86°39'50"E	670	**1996**	September	1233	310	5415
					October	215	59	748
					November	34	22	58
					December	15	11	21
					TOTAL	1497	402	6242
					MEAN	**374.25**	**100.5**	**1560.5**
7	Dudhi Kosi	Rabuwa Bazar 27°16'00"N & 86°39'50"E	670	**1997**	May	17	10	49
					June	169	18	1604
					July	723	348	1421
					August	1694	386	7512
					TOTAL	2603	762	10586
					MEAN	**650.75**	**190.5**	**2646.5**
8	Dudhi Kosi	Rabuwa Bazar 27°16'00"N & 86°39'50"E	670	**1988**	Jan	104	27	251
					May	274	95	804
					October	371	20	3000
					November	96	22	223
					December	273	53	2747
					TOTAL	1118	217	7025
					MEAN	**223.6**	**43.4**	**1405**

9	Dudhi Kosi	Rabuwa Bazar $27^0$16'00"N & $27^0$16'00"N & $86^0$39'50"E	670	**1987**	June	134	62	310
					July	494	320	1450
					August	636	344	1223
					September	274	159	1082
					October	139	80	349
					TOTAL	1677	965	4414
					MEAN	**335.4**	**193**	**882.8**
10	Tamur river	Mulghat $26^0$55'50"N & $86^0$39'50"E	690	**1979**	Jan	479	39	6820
					Feb	142	12	743
					March	125	12	743
					April	528	27	2838
					TOTAL	1274	90	11144
					MEAN	**318.5**	**22.5**	**2786**
11	Tamur river	Mulghat $26^0$55'50"N & $87^0$19'45"E	690	**1980**	Feb	198	44	473
					March	176	34	1227
					April	1337	60	11248
					May	3503	138	9083
					June	8339	2674	16362
					TOTAL	13553	2950	38393
					MEAN	**2710.6**	**590**	**7678.6**

12	Tamur river	Mulghat $26^0$55'50"N & $87^0$19'45"E	690	**1981**	June (only 7 day data)	5786	2200	8910
					July	4882	1490	11200
					August	5790	2900	13900
					September	3963	1800	7510
					TOTAL	20421	8390	41520
					MEAN	**5105.25**	**2097.5**	**10380**
13	Sun Kosi	Kampughat $26^0$52'30"N & $86^0$49'20"E	680	**1984**	May	2039	134	16929
					June	3140	109	7507
					July	6558	2093	14444
					August	5575	2131	12807
					September	10280	2702	37429
					TOTAL	27592	7169	89116
					MEAN	**5518.4**	**1433.8**	**17823.2**
14	Sun Kosi	Kampughat $26^0$52'30"N & $86^0$49'20"E	680	**1974**	May	1590	494	5780
					June	3391	592	17600
					July	1669	683	3630
					August	3861	1092	12092
					September	1406	160	5257
					October	381	54	857
					TOTAL	12298	3075	45216
					MEAN	**2049.67**	**512.5**	**7536**

15	Sun Kosi	Kampughat 26°52'30"N & 86°49'20"E	680	**1975**	April	309	38	2700
					May	1994	64	14900
					June	3259	636	16000
					July	3111	1870	6820
					TOTAL	8673	2608	40420
					MEAN	**2168.25**	**652**	**10105**
16	Tamur river	Mulghat 26°55'50"N & 87°19'45"E	690	**1982**	July	5101	3048	8577
					August (3 months)	2766	2110	3353
					September	8605	1239	41727
					October	2248	284	10661
					TOTAL	18720	6681	64318
					MEAN	**4680**	**1670.25**	**16079.5**
17	Tamur river	Mulghat 26°55'50"N & 87°19'45"E	690	**1987**	September	4345	864	13500
					October	3544	99	19200
					November	760	172	2660
					December	247	73	936
					TOTAL	8896	1208	36296
					MEAN	**2224**	**302**	**9074**

Source : Investigation and Research Division, Kosi Project, Birpur (Supaul), Daily discharge and Silt Observation, Barrage Site.